Nissan Sinai

The Dark Maze

THE DARK MAZE

New and Hidden Insights and Laws of Nature

Nissan Sinai

Senior Editors & Producers: Contento

English Edition: Ariella Goichman

English Edit: BookMasters Group

ISBN-13: 978-1519717627
ISBN-10: 1519717628

International sole distributor: Contento

22 Isserles Street, 6701457, Tel Aviv, Israel

www.ContentoNow.com

Netanel@contento-publishing.com

NISSAN SINAI

THE DARK MAZE

NEW AND HIDDEN INSIGHTS AND LAWS OF NATURE

Table of contents

Contents

INTRODUCTION

In this book, I reveal to you, the readers, some insights and natural laws, new and hidden, that I have discovered in the field of science, particularly in biology. Such insights as physical well-being, mental well-being, the concept of time, the principles of our creation, the conduct of each living organism and more; all these natural laws have thus far eluded us within the mysteries of "The Dark Maze."

How did I discover them?

Listen to my tragic and fascinating story. This story is absolutely true.

MY STORY

This is a story about a legendary thoroughbred. It must be said that horses of this breed aren't very exciting to look at. Not appealing to the eye. Still, as far as domesticated animals are concerned, my love was entirely reserved for them.

My story begins in the year 2001, in one of the rural communities in the south of Israel, the Holy Land, when a newly-born foal of a mare named Maigryaka, and a horse by the name of Bold Arrangement, caught my eye. The parents were fine race horses, but that quality did not pass on to their offspring. This foal was, for a change, exceptional looking. He captured my attention from the first moment I laid eyes on him; a black, ravishing, handsome, strong, noble and muscular foal. His body movement was hypnotizing as he trotted with great energy and long, lean movements. His gallop was elegant and refined, as he seemed to be a creature from another world. I was determined to get him at any cost.

In my mind, I was already toying with ideas for names and found the most appropriate name for this one: Flash Carbonado. (Carbonado meaning a black diamond, which suited his appearance because of his color, but mainly because of the sharp and chiseled angles of his muscles and extraordinary nature). Unfortunately, the foal and his mother were neglected, and kept without food, water, or regular care; all because they were owned by a group of boys with no experience or knowledge.

Later, the foal was left there without his mother or the herd he grew up with, alone and miserable. His sad, extinguished eyes penetrated my heart. He was worried, overwhelmed, and exhausted. Yet, his tragedy had only just begun.

The days went by, but I couldn't stop thinking about that beautiful foal. Finally, when he turned a year old, I got over the obstacles standing in my way and bought him.

This is where the dark marathon began. Because we are taught that animals are emotionless, biological machines, I placed him with some adult horses, in the hopes that they would take him under their wing and strengthen his spirit, and that way he could play with them and develop; and along with the quality food I would provide for him he would be rehabilitated.

To my grave disappointment, the adult horses rejected him immediately. He was heartbroken, and stopped eating almost completely and came down with various illnesses, such as severe flu and diarrhea. In my naivety I believed it was only a matter of time before he got better, since we are all taught that the flu is transmited by a virus, and that it always passes after several days. Unfortunately, his condition got worse and worse, and even the vet's treatment did not work. I have tried different vets over and over, but they all kept injecting him with massive amounts of antibiotics and other substances, to which he only developed acute allergic reactions (with convulsions caused by heart murmurs). His condition continued to worsen.

Back then, in my time of ignorance, no one had connected his illness to his mental state. After a while, I made up my mind to take a drastic step. I wasn't about to give up on such an enchanting creature, so I brought him another foal of the same age to keep him company, and I removed the adult horses from the stable in hopes it would help him recover. But to my astonishment, the young foal, trying to imitate the older horses, attacked him as well, despite his critical condition (or maybe because of it).

How frustrating and degrading that was. From a vivacious foal with great strength and awareness of his own power, Flash turned into a wreck. My heart was broken along with him. I had no one and nowhere left to turn to. He was still under the care of several vets, and each injection he got was only followed by immense and terrifying swelling.

The foal was crashing by the day.

In fact, he became a shadow of his former self. I had separated him from the aggressive foal; I couldn't see any other way. Regretfully, I decided to put him down; which was the custom with distressed animals – no other answer could be found.

In front of me was a poor, shattered creature. His former nobility had become awkwardness, and his diamond-like muscles turned into burnt coal. He became round, greasy, and ugly. His head was dropped; his nose got stuffy, full of liquids, and his body suffered from loose joints. He had a dim cough from the pit of his stomach. His skin had withered and dried; his hair was dry and burnt (even though he was constantly in the shade). He was like a mourner. He didn't even try to swat flies that piled around him as if he were a carcass. Flash became a living-dead creature.

His condition snowballed. No matter what I had stubbornly tried to do, it backfired. For instance, I had tried to spray with fly repellant, but due to his deteriorated condition, his skin reacted harshly, drying up and cracking. I was often obligated to get several mares to keep him company, in hopes he would be interested and to warm his heart. But that only helped for a day or two. I lost all hope.

When the time had come to put him down, I couldn't take that step, as could be expected. My love for him was bigger than the sum of all his illnesses. Besides, there was the enormous amount of prestige and potential embedded within him. But what more could one do in this dismal state – a state of a cruel destiny? Destiny!

To my amazement, several months later I found a small, cute, and quiet horse. A friendly horse that possessed all the qualities I had been looking for in a horse, to keep Flash company. This time, things were indeed looking up: the horse accepted the young foal as his friend, and I could afford to immediately terminate all the vet treatments. Flash did recover from his flu without any medication, after a long and exhausting year; but still, he was never his old self again. It was too late; the damage had already been done. Flash had developed a deep aversion to the stables, the environment, and the entire atmosphere surrounding him, which had left him deeply wounded. Even worse, he developed an aversion to me as well, because of the combined treatment I had provided for him. On the one hand, it was a devoted, results-oriented treatment. On the other hand, it was an intensive treatment, filled with my outbursts of rage, and fueled by frustration and disappointment.

A year passed. The foal matured into a grown horse. When he reached the age of two, to his great dismay, I began training him to strengthen him both mentally and physically (though not for the purpose of racing, due to his condition). As expected, his energy level was at an all time low. For instance, when given the signal to shift from a light trot, he would trip and fall on his hind legs. He was still suffering from gas and constant diarrhea; he would sweat heavily, yawn often and his moaning was worrisome. He would produce those frequent yawns from the depths of his shattered soul. His neighing then sounded almost mute, because his frailty didn't allow him to make any sound. He was forced to inhale air into his lungs a second time, before he could make use of his voice. Only then did his neighing make a sound. I had fallen out with many horseshoers, because of Flash's general condition and I was particularly angry about the fact that his loose joints didn't allow him to even stand on three legs. They lacked any understanding of or empathy toward his condition, and so would hit him any time he lowered his leg. My numerous attempts to explain the causes of his behavior were always met with the thoughtless responses of closed minds.

This excruciating and disappointing marathon, which had been my share since the very first day I purchased him, was gradually leading to my *own* collapse. I had no one to consult with. After all, even the biggest experts in the field could not diagnose his condition. To the best of their knowledge, a fat horse is a healthy horse; and it is up to you to simply train him, build his muscles, and have him run around (being the biological machine that he apparently was).

The days passed, with no solution in sight. I contacted many vets from all across the country and abroad, but they too treated the horse as they would any other animal – as a mere biological machine. They all suggested giving him enhancement vitamins, Glucosamine and Chondroitin for his arthritis, ulcer medications and numerous shots. He was also put through multiple series of blood, urine, and stool tests, in addition to receiving active coals to treat his gas problems. Nothing seemed to be working, and what's worse, some of the treatments almost killed him.

I finally came up with a rehabilitation plan, basically containing the following principles: providing him with a new environment, cutting him off from me, and transferring him to a female caretaker (since I have a deep aversion to the dominating, humiliating, and 'know it all', typical male caretaker). In short: starting over with reasonable and cautious steps, under my guidance, since I am the only one who understands Flash and the depths of his soul. As I anticipated, everybody mocked me. All the "experts" suggested I be more aggressive in my treatment of him and use a whip to push him around. They all ignored his condition and ridiculed my constant claims that the root of this grim condition was entirely mental, and that an aggressive treatment would only make matters worse. Being inexperienced in those matters, I was backed into a corner, and had to take their advice. So I took him out to train him alone with a whip, because of the experts' opinion that the horse was simply lazy, or that his behavior was a result of a genetic problem of some sort.

Again, to my amazement, the horse went into a state of shock and began showing symptoms of Laminitis (a disease that paralyzes his hoofs). That was the last straw. I stood there helpless against a dysfunctional system.

I detached myself from everyone.

During those moments of complete meltdown, changes were beginning to occur within me. I began feeling as though I was able to understand animals and see clearly what was happening within them, both in body and soul, without the use of any unnecessary and invasive means. From that point on I started to develop insights and revelations of things that were unknown to me up until then. That's when the "no choice" insight manifested within me. I discovered that the basic condition for revelations is detaching oneself completely from human conventions and from humanity in general.

I began to see in detail what was happening inside my dearly beloved horse. I could sense that he was suffering from depression, chronic weakness, chronic fatigue, heartbreak and humiliation. He was mourning and generally suffering from a mental breakdown. This caused his body to react with ulcers, an inflammation, numbness, and poor digestion. His skin was brittle and began to crack; his fur looked like patches of wild weed; his joints loosened and were inflamed; his chin swelled up and dried blisters appeared on his tongue. Gradually, he began showing signs of senility: his energy level was almost nonexistent; he seemed to be lacking such vital ingredients as testosterone and dopamine; his organism was engulfed in multiple bodily toxins and he was generally in pain. All these symptoms went completely undetected by the experts and no one appeared to want to hear them.

Once again I tried to explain to all these so-called "experts" that the horse's condition was caused by his mental problems, and hence the treatment should be mental and not conventional. But they all mocked me, claiming that horses do not understand anything about mental issues. "They have no memory, so how can they be suffering from mental disorders?" They laughed at me and continued providing me with their ignorant explanations. I couldn't stand his condition anymore – the increasing yawns, the multiple sighs, and the impossible circumstances all around. It was a mutual collapse, as far as both the horse and I were concerned: a never-ending, downward spiral. I was now facing a tough dilemma. On one hand, I desperately needed to take him out of this environment, which held no hope for him. On the other hand, I knew that whoever took Flash under his care would only cause him to deteriorate further. This was nothing short of a "mental rape." I found myself reaching my ultimate limit. Flash seemed like he was desperately begging me for his life, to free him from the chains of his agonizing soul, and particularly from me and my surroundings. It was in fact a kind of unrequited love: a humiliating one, nonetheless. It was truly a tragedy.

Left with no choice, with harsh reality slapping me in the face, forcing me into it, and with very little common sense, I gave away the most precious thing I have ever had to a man who, even though he was highly experienced in handling horses, had actually tried to hurt Flash a while back.

The man was a professional horseshoer, who had had his eye on Flash for quite a while, and decided to do whatever it took to have him. Once he noticed that Flash had Laminitis disease, he saw an opportunity in the grim situation to carry out his scheme. He had intended to damage the horse's hoofs and temporarily disable him so that my despair from him would grow and I would hand him over. Still, I agreed to turn Flash over to him, because my then poor judgment told me that if he loved him so much, he would take proper care of him. But all my worst fears came true.

The man, as I later discovered, had no knowledge of training horses and had therefore trained the horse with excessive brutality; using a whip, and assuming the horse was just lazy and fat. He would hit Flash for every mistake he made or any immediate disobedience, since he failed to realize the horse was senile, exhausted, passive, and lethargic. He conducted horrific experiments on him to reenergize and awaken him, such as giving him Diesel supplements; or whenever the horse got out of the sprinting device, the fellow would use the whip on his behind to accelerate his leap. At the end of each practice, he would rush him with a whip, to inflate his lungs, to the point where there was no longer any use in trying to explain that the lungs were not a balloon to be inflated, and that there are reasonable methods to start over with. In addition, he also fed him moldy food and would starve him before each race to keep him light on his feet.

Furthermore, he sprayed Flash with bug spray, which wasn't at all meant for horses, but rather for agriculture, since it contained such materials as sticky sugar and other revolting smelling substances. All around me, I was hearing many ridiculous theories (such as: "He loves the whip", and "We need to 'pump' him up," leaving Flash alone with his screams, inflating his lungs).

I was fully aware, down to the last detail, how my beloved horse was being abused, and how his already crushed soul was "being raped." During one of my visits I discovered that his upper lip was partially cut off, with bleeding and pus on the wound – a result of abuse with a rope's loop around the muzzle, some part of a twisted therapy. A tragedy.

I couldn't find a way out.
Following two failed attempts, I brought Flash back to me, after he had suffered from "cabin phobia," and refused to come into my cabin. However, the caretaker refused to return him to me, perhaps as a means of blackmail. But then again, it seemed as though he was becoming attached to the horse. Still, it was a frightening and devastating experience, as far as I was concerned, and I was about to call it quits, but backed off at the last moment, realizing I had a genuine fear that the horse would be harmed or

even poisoned. I was convinced that some devil had made a target out of me, but giving up was not an option.

I tried to rehabilitate him again, but soon after he was stolen from my home by two low-lives from a distant area of tents. The previous attempt of blackmail had motivated them to try it for themselves. They rode him together and abused him, which is, in fact, the method of forcing the owners to pay ransom. When I found him, he was exhausted, shocked, and passive; with his tongue all ripped and hanging out. This time, Flash completely stopped eating and drinking. Whenever he tried to drink, the water spilled right out of his mouth. He couldn't swallow and eventually collapsed. The stench coming from the stables reflected the emotional state he was in.

That's where I made the most fatal mistake of my life: I sold him. The horrific scenario had repeated itself. Other than the overall tragedy, the sadistic previous caretaker continued taking part in his care, despite my unequivocal warnings and instructions. My voice was silenced, the skies fell on me, and I was sinking deeper into the depths of my own ocean. I was in a dangerous state, believing that I was still battling the devil. This time, I couldn't take Flash back.

Flash continued to compete in "mechanical" methods and even beat the country's finest imported horses. Unfortunately, no one noticed his weakened body language, his constant yawns, the bald-spotted fur-less skin, the inside of his rotten hoofs, the endless diarrhea, the swollen stomach, the loose joints, his dried throat and bloated chin. On top of everything else, he would stop at any moment, holding his tail up in a desperate attempt to pass gas. Despite his "ridiculous" victory, his achievements were far below his natural ability, and even after winning a race, with his stomach rumbling up and down, he was photographed for a newspaper article in a midst of a deep yawn. After that, there was no one left to speak to. But even this odd photo didn't make anyone wonder.

People who lacked minimal understanding of horses passed Flash around as if he were a cheap object. On a visit once, I found him alone. Yes, Flash with his tattered soul, who had been suffering from severe abandonment issues, was left alone. He seemed pathetic in his deteriorated state, lying down with a sore stomach next to the moldy, smelly hay.

When I finally voiced my concerns, one of his caretakers simply said in response: "Why are you so worried? Horses aren't people. I'll do my best to buy him some food the first chance I get and fatten him up within a week."

I ran out of words. I ran out of options. All that I was thinking and trying to explain fell on deaf ears.

I sank further to a point where humans were beginning to seem like a tragic mistake of evolution to me.

Here is where I pause for a moment to go back to a rare occasion in which Flash showed a fraction of his lost abilities. That was before I had given him away to the first guy. That is, before Flash had even participated in any race. I shared this experience with no one since it was clear to me that nobody would believe my story – and that they would even mock me.

This is the tale of true events: during one of my longing-filled and hopeful attempts of training Flash, I had him compete in an unofficial race with a veteran racing pro – a mare called Furaja – near the town where I was living. I didn't use a whip, and made no prior preparations. I only counted on Flash's competitive nature. We started running, but Furaja was eight lengths ahead of us. Flash didn't even seem to want to compete. It was his first time; he had never even had a taste of racing. All he seemed to want was just to be around the mare. It was a classic case of the miserable boy fascinated with the female guest of honor. As soon as he realized she was advancing and slipping out of his sight, he launched a gallop, such as I had never seen before. Within seconds he was right up on her tail, aggressively biting her behind, to try and trample her. I had tried my hardest to pull him away from her by pulling the reins tightly, but Flash reacted by going into a frenzy, bypassing her from the left and ignoring the stop signal. My massive tugs on his reins were futile, as he shortly left her far behind, despite his condition! At that moment I was reminded of his appearance when I first saw him – a unique, one-of-a-kind creature.

That moment never repeated itself, and his hidden potential was buried forever…

Back to my story:

I was alone in my battle, full of expectations from a horse, who, in my mind was supposed to be a "Super Horse." I had hoped that professionals from the field of biology sciences would possess the magic that would cure him of his misery. But no one was willing to listen or understand what was going on, and my cries fell on deaf ears, which only resulted in my roaring silence. Flash was now a hunchback, bent over and worn-out. He had lost all his virtues, charm, and beauty. Yes, man had won.

Flash is a stallion now, to a mare by the name of Winter Jasmine, whom I purchased specifically for this purpose, hoping to produce a future generation that would carry out his lost uniqueness. I did this out of spite to all those who had mocked his theoretical abilities, denying the notion that he could function as a stallion, and all those who had even declined, with mockery and contempt, my proposal to breed Flash with their own mares. I had only the hope that his excellent genes would pass to future generations.

The first female foal

Yes, I was ready for all disasters to strike. The situation had gotten out of hand, due to an accumulation of further events I am forbidden to share. Darkness fell upon my world. The slope was getting even more slippery; my body was falling apart. Fireworks that were lit near the stables frightened the mare, causing her to go into trauma and suffer a miscarriage. It was only a year later that the first female foal was delivered, but she was rejected by her mother because of the mother's paranoia, her lack of experience and her aversion to the stables. She began to abuse the poor, innocent foal, and hurt her badly, both mentally and physically. There was no way to help the foal, because the mother would have a fit any time anyone came near her. It was a diabolical and malicious black hole, which sucked and trampled unsuspecting victims into it. The female foal was born massive and amazing, but had turned over time, due to her failing mental state, more and more abnormal.

When her younger brother was born, my heart was completely broken. I had rehabilitated her to the best of my abilities, and I avoided contacting any so-called "experts", since I could play the chorus in my head already: "It's genetic" and those sorts of things. The best way to cure her was to disconnect her from her monstrous (and yet beloved) mother. To this very day I'm still looking for new owners for her, who have a head on their shoulders and will not treat the foal as a commercial object. I was made ill listening to such cruel sayings as: "Why do you care who buys her?" "Take the money and get rid of her already," and "They should make her labor with a cart and a whip – she's not yours anymore anyway." This twisted behavior had reinvigorated my urge to explore the "human creature" all over again.

In time, I have successfully rehabilitated her mother. She (Winter Jasmin) gave birth to two other foals, with Flash being the father, and she's now a good, stable, and experienced mother. Although they are much "improved" compared to their clumsy mother, unfortunately I know that the incredible creature (flash) is, indeed, incredible and therefore does not multiply according to expectations.

That is the essence of my tragic story, told in a nutshell, so as not to tire you readers out there. And it's still far from over.

After having been through such impossible experiences, I was swept over by an obsessive urge to step out of the robotic conventions and discover hidden, unknown secrets: to solve them, and by doing so to complete this so-called puzzle of our existence. It's now or never.

In order to discover hidden revelations, one must pass through all cycles of the inferno of emotional distresses, experience impossible situations, and come out of the other end. One must roll on from a sadistic, fanatically religious, labor-forced childhood and onto "my story," like a snowball falling into the dark ravine, with one crisis following another.

To survive I had to acquire different and hidden techniques, which I acquired by the "no choice" instinct (which will be explained further later on in the book). Since one couldn't live under such conditions as I have lived, this was an extreme situation where the "no choice" element urged me to open up my narrow and fixed horizons.

Once you do expand your horizons, you can see clearly

what you couldn't have seen before. There were many compelling reasons to rid myself of this twisted reality and miserable life. But on the other hand, there were other reasons that forced me to stay alive at all costs. As this is indeed a short account of events. I have no intention of rummaging through the dark marathon of the past, because we have little time and a lot to get done.

THE DARK MAZE

For you, the reader, to fully understand the essence of this book, I ask you to prepare yourself mentally to step out of the conventions and common beliefs that have thus far been deeply rooted in our society. A man is but a slave to the copying and mimicking of others: a slave to his set of habits, beliefs, fixations, prides, his acquired knowledge, and so forth.

The laws of nature, which I have begun to reveal, are nothing like the conventions we were taught and raised with.
This book is dedicated to the lost man.

Note: The insights in this book are only recommendations.

The use of this book is conditioned by an absolute under-standing of its described principle.

Intelligent readers: don't wait for the experts to read, understand, approve, and distribute the book, because they're already in the dark maze. Their information has already been set, which is based on faulty founda-tions acquired from their predecessors whose predeces-sors acquired it from their predecessors, etc... etc...

Our brains, our bodies, and ourselves
How does our body operate?

Our bodies, including our brains, function with glorious accuracy, for better or worse: by unknown, internal, infinite and individual sensations of the equally infinite and individual "self." These sensations occur through infinite individual interpretations of the infinite "self" against infinite situations, whether they are real or imaginary.

This principle applies to the entire animal kingdom.

Every organ (including the brain) and every cell operates by the infinite and determined nature of this principle. No blink of an eye is made by a spontaneous decision of the brain.

Yes! That's what we have missed: that slippery "self" that is unseen and intangible. Unlike what I have thought up until writing this book, the brain is not insightful or understanding and it cannot distinguish right from wrong, reality from imagination, and past from future and present. The brain doesn't care for our best or worst interests; it doesn't give us instructions like, "Walk right or left"; it doesn't instruct us to love or hate, believe or disbelieve; and it doesn't resolve our hardships in life. The brain doesn't have a "mind of its own." The brain is not what we have thought it to be.

In fact, our enormous and wasted brain would have been useless, if not for our separate "selves."

The brain takes orders from the endless "individual self."

The brain is the extension that connects the "self" to the body.

The "self" is the ultimate owner of the body and brain. The brain can be compared to a super computer. Its user (the "self") chooses whether to use it or not, and makes the most of it – or not at all. It is also the one to decide how much, and in which quality and efficiency to use it. Yes, we are in fact using our brain and not the other way around. I am the one writing this book, not my brain. You are the one reading it and not your brain. "We" is the "self" of every living creature that thinks with its brain. Phrases about the "self" like: "I", "you", "for me", "for you", "I will do", "I don't feel like it", "my heart broke", "he's eating himself", "you're giving me a headache", "I'm tired of...", "I'm glad", "I'm feeling butterflies in my stomach," "I think such and such," "She changed her mind," and more, are uttered without our awareness.

These insights, along with others, which will be presented later on in this book, are supposed to rock a man's consciousness and shatter conventional world views.
In this book I shall explain about the operation of the body and brain; I will discuss in length physical and mental illnesses; and I will add new and hidden laws of nature.

I would like to share my knowledge for the welfare of mankind, animals, and the world in general. For that reason, I am asking professionals from the field of science, psychology, and other experts, who share my hopes and ambitions for a better tomorrow, to focus on the principles of this book, because there's a lot of work ahead. I hope that with their cooperation, the changes in our lives, health, brain functions, and our world will become a reality. My intention is to take everything that was rooted within us thus far and shift it by 180 degrees.

Every mystery has its solution. The fact that none was found doesn't mean it's not there. We shouldn't bury our heads in the sand, and turn to alternative solutions and beliefs as a reaction to a lack of solutions. That's because a wrong solution can be entwined in a man's consciousness and pass from generation to generation – even if it sounds brilliant (like religion).

The Principle of the Brain's Conduct

In one neurological study, a scientist attached electrodes to the brain of a volunteer, connecting them to a computer. The scientist asked the volunteer which of the two light bulbs he preferred: the red one or the green one. The computer produced the volunteer's preference even before he had replied. The scientist thus concluded that the brain knows the volunteer's preference even before he knows it and reacts to it.

That's where he got it all wrong.

My obvious question would be: before *whose* knowledge and reaction exactly?

In this case, it is the tested "self," and its subconscious early response, as well as its late and conscious response that play the main role. As I have said, the answer is made out of personal commentary, i.e., his own taste and choice of the light bulb's color. The responses are in fact individual sensations caused by individual interpretations of the tested "self." The anticipation may be for another second, minute, hour, day, and so on.

For instance, if we sorely pinch our tested volunteer, his sensations will be that of pain. The body will operate and react to the painful sensation of the body owner, who is the "self." The tested "self" can respond to the pain by:
Making an immediate and instinctive "Ouch" sound.
Restraining itself and responding later on.
Not reacting at all.

However, the painful sensation arrives immediately

following the pinching. But since the brain cannot differentiate between reality and imagination, a situation occurs in which the electrodes respond prior to the pinching sensation. In this situation, the "self" feels the upcoming pain before its actual realization. The brain is obedient to the endless "self." The responses are in fact individual sensations according to individual interpretations by the tested "self."

THE GENESIS THEORY

There's a common belief that there is a creator and we are its creation. If that is the case, then who created the creator? (See **"Concept of Time"** chapter). In order for our brain to manage us and our bodies, supposedly by its "own mind," we need a creator for the brain. Yes! We need a primal entity that made our bodies and minds before it created us. **Yet there was no entity of this kind, nor does it exist today!** Say, theoretically speaking, that there was one. Then the obvious conclusion would be that the living creation is a robot, dominated by its maker. And if the living creation is a robot, then who operates it? That's where we go back to the "self."

What is the process of "fact and no choice?"

Space = nothing

The temperature in space is minus 280 degrees Celsius. Space contains material and movement, such as dust, gas, and so on. Dust absorbs gas, gas absorbs dust, and gas absorbs gas. Then, more and more dust and gas are annexed to this crystal. When the crystal reaches a certain size, gravity is created by the accumulation of the process of "no choice." The gravity in the crystal attracts more and more substances from its environment by reaction, until becoming a star. Inside such a star, mass is created. This mass creates chemistry. The chemistry creates heat. Heat creates energy. This energy provokes endless activities and results. One of these results is the cell (meaning, the scientific convention that determines that it is impossible to generate energy without a source of available energy is wrong). Trees, for example, don't have a pump to pull water all the way up to their top.

Every phenomenon or chemical reaction is caused by the "no choice" process. Each cell is made under certain conditions. The cell knowingly multiplies itself again and again, creating a "lump" of cells. This lump of cells cries out for energy, so it will go on existing. That's where the wheel of the "no choice" process starts spinning. This mass of cells organizes itself so it will absorb energy in the most efficient way (sunbeams, food, and oxygen). As time goes by, primitive organs are made to better themselves. Because there is a shortage of food in nature, there's a war for survival. That causes the creation of a primitive brain, with only a subconscious mind to operate the survival organs in a more efficient way in the "no choice" situation.

And this "no choice" is the uninhibited "*I am my own maker,*" which the brain has created in the form of the conscious, the subconscious, the body, the organs, and the senses.

A nice example for this theory is mold. In the beginning, there was a primitive creature, created in the appropriate conditions. Later, following the "no choice" theory, its arms were stretched like cotton balls to absorb the moisture, which also contains such energy as minerals.
From that point on, the "self" is the uninhibited "no choice," longing for continuity.

Every creature cries out for energy and continuity, hence the "no choice" instinct yearning to send its roots to life. This principle had, thus far, slipped out of our consciousness, thinking that a living organism is charged by and functions on food alone, and that his steps are programmed by the "boss" – the brain. May I remind you that without the "*I am my own maker*" (the "no choice"), the body wouldn't have asked for food on its own accord. A body that is not governed by anything doesn't have needs, desires, or aims of its own. And generally, there was no body without an owner to begin with.

If that's the case, then every living creature creates itself in its own endless and subconscious ways, as part of the "no choice" drill. This process comes to life instinctively, when the "no choice" factor – that is, the "self" cries out for energy, life, continuity, and self-division. That is the *Ultimate "I am my own maker"* of each individual. We'll term it simply the "self."

The structure of the "self"

We were not created by a primal, pre-historic creature or by any particular entity. For that reason, the conclusion is that we are **all alone**! Each and every single one of us stands in their own right. The brain does not take care of us; neither does evolution nor any type of entity. For instance, when it rains, it does so according to certain conditions. It doesn't rain particularly for the animals or plants, but each living creature makes use of it through various methods in the "no choice" process.

SUBSTANCE AND LIFE

Since when is substance alive? Can a single cell be a living thing? And at which point following the insemination could a growing fetus be called "alive"?

Flammable substances don't wait for fire to be ignited, but when there's a spark, the fire feeds upon the flammable substance, setting it aflame. The fire sucks on energy, which is the flammable substance, and oxygen. Fire takes useful energy it so desperately craves (and not due to any intent, awareness, or will). As there is more flammable substance around, the fire spreads even farther. Fire is not a part of the existing substance of planet Earth. It is an inevitable result of a random occurrence.
It's the "no choice" phenomenon.

We can also take acid, for example. Acid doesn't wait to burn the surface of a different substance. But when any substance interacts with acid, it burns it. It feeds on the substance, causing a chemical reaction.

A dry seed of a "dormant" plant doesn't wait for the rain. But when it does rain, it feeds the seed. The seed stretches its "arms" to cling to life. Under certain conditions a cell is made. When the right energy appears (such as food, rays of sun, and oxygen) around the cell that is crying out for energy, it then takes it and feeds upon it. That is the spark and those are its consequences. Is it alive? Depends on whom you're asking.

Life begins once you cry out for life. For example, the sperm cell is not yet alive, but it strives for life. It is created to actively swim by the "*Ultimate I am my own maker*" urge, which is part of the "no choice" process.

So what is "alive"?

Life, by definition, belongs to a creature with an awareness of its life, which exists in the center of life and is hopeful. For instance, domesticated animals that are hopeless, due to many reasons, should be put down to save them from their miserable, frozen lives. It's better to castrate them than to condemn them to an unhappy life, as well as the fact that those living creatures didn't choose to be born and live.

BODY MANAGEMENT

Like humans, plants too have an active sexual reproductive system. But that's where the similarity ends. This is not the case for living creatures of consciousness. The fact that many books have been, and will continue to be, published on the issue of sex, and sexual successes and defeats, indicates that there is a missing link.

This link is the brain, which doesn't differentiate between good and evil, which is not human, nor does it understand or care for our sexual welfare. It doesn't care to fix various problems and complexities, and it doesn't care for our downfalls or even our successes. It doesn't guide us by usage instructions. And again we are alone. In light of this, we should add psychology and sexual education studies to our educational programs, even at a very early age, so that we don't get lost. The brain must not, under any circumstances, be integrated into this, because **the brain is not relevant, lacks usage and guidance directions, and does not care to fix anything. It is not emotionally attracted or averted. It doesn't have its own taste or smell. It does not fear, cry, or experience happiness. It does not love nor hate; it is neither sad nor depressed; it isn't lazy or hardworking; it cannot tell the past from the present or future, or imagination from reality; and it has no preferences of its own whatsoever.**

I repeat with emphasis: the body doesn't live and grow

for granted. The brain doesn't have an awareness of its own, because the Ultimate *I' am my own maker"* urge created the body as well as the brain. The brain doesn't run our lives, doesn't think, or worry for us. The brain is only an obedient organ of the endless "self." This new insight takes getting used to. Each living creature is born completely helpless, to a life filled with survival challenges. The body doesn't carry itself around, but the endless *"Ultimate I am my own maker"* does. Because of that, the life of the "self" is mentally tasking, and the body carries itself subconsciously, yet with lavish accuracy according to the *"individual* self's" sensations. According to this insight, life is like a **"quicksand swamp"** that conducts itself indefinitely. This "quicksand swamp's" hidden principles will be explained throughout this book. Thus we learn that physical and mental health are not to be taken for granted, because it is the "self" that dictates and manages the body and its entire being. Every living creature survives by its individual, diverse, and unknown ways. **The instincts lead us to be more and more victorious as an expression of striving to come up from the "quicksand swamp," even when the struggle has nothing directly to do with survival.** And that is a hidden secret you will now discover.

Why do we have a constant urge to always drive relatively

faster and pass the car that is in front of us? After all, when we have passed the first car, there will always be another car ahead. Why do we have this urge to make it first to the top? Why do we strive for victory, even when we are not at war? For example, when we are playing a game and try to win, knowing that we don't have to win in order to survive, or when someone curses or offends us, why do we push back, and respond to insults and curse words? We again go back to what I have mentioned before: our body is managed by the endless structure of our "selves," and if life is like a "quicksand swamp" (of which we are at rock bottom), then the body works the same way! **Our body is inefficient, to say the least.** An inefficient body acts and carries itself exactly like the drowning "self" – it suffers from inflammations, arthritis, a sickly digestive system, bad skin, lack of energy, and so on. Its organs exude harmful substances, while positive substances like testosterone and dopamine get lost and become irrelevant, much like the sensation of drowning. We fight by harming others to get ourselves out of this imaginary swamp, and struggle to stabilize the inferior "self," and raise it higher and higher. As we strive to climb upwards, our bodies match our sensation and vice versa.

All this materializes without our awareness,

instinctively, according to our demand of energy, life, and continuity.

Now, you readers decide how efficient or inefficient we are. I remind you of my tale of the poor, inferior horse. What happens when an individual feels victorious? The structure of his "self" changes according to his uplifted sensation, and the quality of his body therefore improves. The good substances improve and the harmful substances are downsized. Take for example, a man cursing or insulting another man to feel superior. No entity created the anger we feel for, or because of, something. An individual gets angry due to personal interpretations of specific situations. It's a sensation similar to carrying baggage and being unable to unload it. Once you release that anger, you are able to conduct yourself freely. You feel as if you're "above the surface of the swamp." But if there is no release, then the sensation remains within your body, which will carry itself in accordance with the body owner's sensation – that is the "*individual self.*" This sensation can manifest itself through ailments like ulcers, heart distress, dropped shoulders, weakened muscles, aching body language, and much more.

The same is true for the aggressive foal that attacked Flash. His sensation was overwhelming, while Flash's sensation became weakened according to the "Scales Principle." And so, Flash suffered, among other ailments, from ulcers, because he was "eating" himself up. We should keep in mind that it's neither the body nor the brain that create this drama. It is the foal's "*I am my own maker*" and Flash's "*I am my own maker*" that manifest themselves through the brain. In this situation there is no mechanism or entity to pull us out of the "quicksand swamp" other than ourselves. As a living creature goes under, deeper and deeper, it will have more trouble striving upwards again, because the body's senses and nature have become inferior and downgraded in accordance with the drowning depth. On the other hand, there is the winner that savors this situation with great pleasure. It's a war for survival.

At times of trauma and collapse, taking some remedy or potion to numb the senses could actually cause some well-being, to a certain extent.

SENSES

The sense of hearing

The sense of hearing developed by a vocal interpretation of the surrounding events. For example, stepping on dry leaves creates a squeaking vibration that is translated into hearing. The "self" listens to the sounds and is mindful of its surroundings (like a predator ambushing the prey). The "self" also expresses itself with voices like grunting and screaming, or cries for help that echo in the body and its hearing interpretation.

That's how the hearing organs have developed.

The sense of pain

The sense of pain was created by the "*Ultimate I am my own maker*," which is anxious about its own body. The body is constructed and built according to the "*Ultimate I am my own maker*". When threats arise to the wellness of the body (like wounds or predators), pain is the instinctive reaction to the hurtful sensation. Let's say we anesthetize a person's eye by giving him something for the pain. Won't he react to the threat approaching his eye (like a mosquito that might sting) despite the fact that he knows he won't feel any pain? The answer is quite clear. Instinctively, the subject of this experiment will protect the eye of the "*I am my own maker*." It's a sensation that says, "My eye hurts" even before any pain is sensed.

We may thus conclude that pain is the evolutionary reaction made by the "*I am my own maker*." But pain doesn't always protect us. In fact, it even tricks us at times since there's no logic or symmetry in its design, such as in cases of chronic back pain or headaches.

Conclusion: Pain can be controlled to some degree by altering our interpretation of the situation (explanations will be provided later on in the book).

The sense of sight
The sense of sight developed in parallel to the "scenery viewing self" and prior to the development of the eyes. Suppose we cover a monkey's eyes. When that monkey climbs a tree, he bumps into branches, he feels them, holds and picks fruit of various shapes and sizes (some of which are poisonous and inedible while others are edible), and he smells and reviews the vibrant and threatening surroundings. As mentioned, the tree's image is created in his imagination without the help of his eyes. This instinct helped develop the eyes through the "self," which struggles to see and is invisible.

Sense of smell and sense of taste
The sense of smell developed in a simpler and more basic way.

When a living organism exhales and inhales oxygen, it also inhales smell molecules, dust, disposable chemical substances and so on. That's how the sense of smell developed.

The sense of smell led to the development of the sense of taste, due to the existence of burning and poisonous foods, as well as unripe fruit that poisoned the body. The "self" senses the poisonous factor through the sense of pain in addition to the sense of smell. That's how the sense of taste was developed.

To summarize: It's the "self" that uses its biological organs by individual sensations, and in accordance with personal interpretations (it's not the brain).

THE CONSCIOUSNESS

The consciousness and the conscious actions are limited. One can't build a human body and its organs with the consciousness – **this is done by the infinite subconscious**. At the beginning of the evolutionary process, a primitive brain was formed in the subconscious structure to use as an instinct for efficiency, absorbing of energy, and survival. Later, the consciousness was made to use as an instinctive force to improve the survival rate through conscious activities, according to the ability of each living organism. We are talking of such activities as picking a ripe fruit by a certain color (red instead of green, for instance), turning right, left, or backwards, or climbing a tree. All of these are conscious actions that the conscious "self" interprets by the individual information acquired within the subconscious.

Why is the consciousness limited?

The consciousness is limited because the conscious "self" is a kind of specific **scanner,** which we use in a particular way. Take looking, for example: try looking at scenery of some trees and focus on one of them. The view narrows as a result of this, since the focus is directed at the tree you are looking at. Meaning that at that period of time, the view isn't relevant. But when you draw your gaze away from the tree, the view returns with greater detail to the consciousness.

The same is true when we look at a particular leaf on that tree – the focus on the tree as a whole becomes less relevant. In other words, focusing on the leaf will make the total view of the landscape disappear from our consciousness.

Another example is trying to remember a certain individual – let's say a girl. Trying to recall that person, one will obviously see a general image of her. But if we focus on her face, we could scan her different parts such as her nose, eyes, lips, face, and hair. In the meantime, our peripheral sight of her disappears because the consciousness can only scan specific things.

Try to focus on two images or two objects at the same time. It's naturally impossible. If we attempt to do that, we will find that the memory bounces from one image to another.

The same goes for hearing. Try listening to the news and reading a book at the same time. You will find that you're scanning some of the news and some of the book: a bit from here and a bit from there, at an enormous, subconscious, inefficient speed, less efficient than usual.

Try to think of Antarctica and the Mojave Desert at the same split second.
That's why we are either right or left handed (among other reasons). We cannot operate two hands simultaneously and do it as efficiently as using one hand. Ironically, the brain and the body are able to do so. Ideally there should be no limitation, so it is actually the "self" that is limited. Try writing down numbers with both hands at the same time. An even harder task is trying to write "1-2-3" with your right hand and "A-B-C" with your left hand at the same time.

You can also try and read a line from a book, trying to focus on all of the words in the line at once. You can't do that, of course. We tend to scan each word separately and even letter by letter, adding them up at the speed of a flash of a light, without being aware of it. That's the conscious "self" and its structure. The reasons for this are simple: A). The body is managed by the subconscious in infinity. B). The infinite data that we accumulate throughout our lives doesn't allow us to perform abstractly or inefficiently. Having no other choice, whenever we perform a certain task we become aware and observational of everything else. For instance, we see an overall image of an audience and continue performing the same actions that we have intended to perform. If we focus on one man in the crowd, we will reduce the efficiency of such functions as remembering, executing, saving, thinking about past, present and future, planning future moves, watching out and trying not to make any mistakes, learning a lesson, fixing, bettering ourselves, and focusing.

We derive information from the subconscious database as much as possible, according to the ability of the "self." Contrary to the consciousness, the subconscious conducts itself infinitely and abides by (without any scan) the subjective and the infinite sensations of the "individual self" simultaneously. That is, it acts according to the information that is gathered, whether it is right or wrong, justified or unjustified, accurate or misleading, or even imaginary. The subconscious doesn't have its own wisdom or will, and it doesn't solve riddles in its own right.

In fact, animals (much like us) "talk" without words and with constant thoughts. The interpretations of these thoughts are translated into individual, infinite sensations. The difficulty is to express them consciously. A well-known fact is that animals have very limited means of expression. Humans, however, have more developed abilities of expression.

THE SUBCONSCIOUS

If we were to choose a title for the brain, it would have to be the honorary title of **Wasted**. Seeing as the brain is a sort of complex, biological computer, such titles as "smart" or "stupid" are irrelevant. But the title, **"very complex biological super computer,"** is quite relevant!

The brain does not differentiate between good and evil; past, present or future; imagination and reality; or dreams from films. Since the brain relies on the "self" in the evolutionary process, in order to manage the body in the "no choice" method, the infinite structure of the "self" must be in charge of managing our bodies in an exact and accurate way, for better or worse. The subconscious is, in fact, infinite information, decoded by the infinite "self" according to the perception, understanding, and interpretation provided individually by each living creature. **Accordingly, the right or wrong information is sealed within the subconscious**. Infinite, individual information is an inseparable part of the striving and longing for energy, life, continuity, and reproduction. And it guides the construction and management of the body for better or worse, without the execution of any logic, symmetry or will, and without following any conventions. The eyes and ears don't tell us what we have seen or heard, since the brain doesn't interpret these actions on its own. Mainly **it doesn't think for us,** but rather it is the infinite, individual "self" that interprets the sights, sounds, and thoughts using the brain. The same is true for the other senses.

A remote tribe, for example, suffers from one crisis after another in the shape of natural disasters, droughts, plagues, or hunger. The head of the tribe doesn't know what to do and his individual "self" interprets these events (with his non-understanding and non-interpretive brain) as though the thunder and lightning bolts are expressions of the wrath of the gods. Therefore, says the head of the tribe himself, we should offer them a sacrifice to appease them. He offers a sacrifice and nothing comes of it. He offers yet another sacrifice and another, and yet another – again, nothing happens. But once he offers the sacrifice for a fifth time, the rain starts to pour, and so, according to his understanding, the so-called "wrath of the gods" vanishes. From that point on, the gods become more tangible, concrete and legitimate, possessing super powers. Therefore, in his mind, this tradition should be carried out. The parents pass it to their children, grandchildren, and great grandchildren. That's how a set of beliefs travels among the generations and is rooted in our lives.

In the same manner, the brain doesn't tell the tribe to limit its birth rate according to the existing natural resources. Fate and hunger are, as far as the head of the tribe is concerned, manipulated by the gods, although there are still crowdedness, hunger, and plagues. That's how the human race advances toward self-eradication, as well as the destruction of this amazing planet. Again, this is the interpretation of the "self" without the interference of the brain.

Another example is this: man hunts the marvelous tiger because he believes that a soup, made out of the tiger's genitals, strengthens the libido. In another place on the planet, they hunt the mighty rhino because people believe that the horns are responsible for enhancing the libido. And that's how it goes on until the animals are extinct. We should say that this is nothing but a belief that is certainly not rooted in any scientific fact.

PAST, PRESENT, AND FUTURE

Since it is obvious by now that we shouldn't blame the brain for our physical and mental disabilities, **so it is redundant or irrelevant, that we should praise it for our success and health**. The brain doesn't differentiate between the past, the present, and the future. The future is not relevant and it has yet to arrive. When we think of, or remember, the good or bad past, our body reacts according to the senses of that particular time. In a traumatic shock, the individual relives his traumatic past in the present. He keeps focusing on it, convinced that that is the current circumstance of his surroundings. In other words, the past is the same as the present for him.

Another example is of someone thinking of their previous successful sexual experience. At that moment, that experience is as tangible to the individual "self" and is as sexually stimulating as though it were happening in the present.

The same goes for the future. Since our bodies are managed and directed by the determining "*self*," our responses will be in accordance. When a person thinks about dramatic events that might happen in the future, the body reacts at present according to its thoughts on the future. The individual's heart rate will increase because of the traumatic experience he is certain will take place in the future. As far as he's concerned, the future is tangible at present. Or when a man is focused on a date with his lover in the near future, imagining the great sexual intercourse they will have, his body reacts, depending on his focus on the event and his sensation of sexual arousal.

Evolution didn't bother fixing the distortions in the rules of survival; neither did the brain or any other entity (unlike the laws of science). For example, just when a man approaches a trying moment, or a dramatic and scary situation, he starts to feel his mouth drying, his body shakes, and he might even experience general paralysis. This is a legitimate sensation, since according to his subjective sensations, at that moment he is experiencing fear.

On the contrary, a different individual preparing for a similar situation would not necessarily suffer from a dry throat or a shaky body. Instead, he might feel he is victorious and conquering, even before the event ever took place. **That is in accordance with his sensations, manifested by his individual interpretation of the situation.**

When we are focused on the present, the body reacts and conducts itself according to our infinite current sensations, whether they are positive or negative. Our subconscious is the place where the endless mixture of our individual sensations occurs. Those are managed by our thoughts and interpretations regarding the past, present, and the future. These thoughts and interpretations also manifest in our sensations. Our body is managed by this principle with great accuracy, in a manner that isn't necessarily reasonable or constructed. Meaning that **it depends on what the "self" is more focused on, subconsciously.** When a man is sick, his body reacts and conducts itself according to the sensations of the "self," which may be manifested in sweat. Or if he is having an erotic or traumatic dream, it may be manifested by strong heartbeats, but **the brain can't differentiate reality from dreams, and it doesn't operate upon our will or defiance.** We cannot dream in a custom-made fashion.

FEAR

Fear isn't created for a certain goal. There never was nor will be any entity that created our sense of fear in us to protect us. The time has come to stop this convention.

Fear is in fact an individual sensation of the "self," formed according to its interpretation of a certain situation. It could be a situation that took place in the past, or that is happening at present, or that is about to happen in the future. This feeling comes from our longing for life.

Fear tricks us rather than protects us. When the "money time" approaches, a dry mouth or trepidation can fail us. Fear is paralyzing us, confusing our senses, weakening our bodies and sometimes sealing our fate (like it does with animals).

Walking on a thick plank (when it is placed on the ground) is possible to an extent. But if that same plank is placed fifty meters above ground, the performance is most likely to go wrong according to the fearful sensation of the "self," through its reaction to the frightening, upcoming situation in which the person may fall.

Flight that is accompanied by fear reduces the body's efficiency. A fearful and unwilling assault only weakens the body. That is, of course, in accordance with the sensations. However, fleeing with confidence and without any fears, stops, or breaks, is more efficient, causing the body to react better.

Some people are afraid of cockroaches. Some are frightened by a certain scene in a film and others by nightmares. That is the "self" that interprets things individually and instinctively, by its definition of a real or imaginary event, which took or didn't take place in the past, present, or future. That is done without a guiding hand that leads us, making sure we internalize the fear. That, in itself, is another waste.

We might say that if it wasn't for the fear, we would live our lives according to fearless considerations. For instance, we would avoid confronting a lion or an elephant, not because of our fear, but because we are aware, through rational consideration, that there's no chance of winning this dangerous confrontation. We would prevent ourselves from jumping from the tenth floor just because the "self" would have interpreted the results in advance, with the use of the brain, and not because of any fear. The brain isn't afraid. The brain doesn't direct us to fear or alternatively to be brave in order to survive. It's the infinite "individual self" that longs for life and is anxious about it.

In conclusion: Fear is redundant. When danger approaches, we have awareness of the risks. Again, I repeat – we should shake off the conventions we grew up with. Mankind should go through a change and revitalization.

Evolution doesn't care for us nor does it decide to be as it is. It is constructed on the false perception that evolution is responsible for every living creature evolving, but that isn't the case. Evolution isn't an entity. The correct assumption would be that certain developments were created **throughout** evolution.

PLACEBO

The "self" determines the procedures conducted by the brain and the body. Physical and mental ailments legitimately and accurately evolve according to individual interpretations of the "self" and its lack of resistance, as it is in a state of "drowning in the quicksand swamp." It is twisted, lacks vibrant life, inhibited, and unable to lead.

Every illness has its beginning and once it develops, **it starts to snowball.** This happens because the "self" feels the disease, focuses on it, and pushes it **without its awareness** into the abyss. Frustration builds and leads to more frustration, which in turn causes a collapse (this explanation also relates to mental health issues) and so on.

When a man has nowhere left to turn, he needs a life jacket. Every man should have his own, individual life jacket that fits him and his current situation: a kind word, some support, sex, success, changes, love, a smile, encouragement, or a winning hug. Alternatively, it could be the power of a prayer for a believing man – as long as the believer is convinced that his prayer has been received, and that he senses the situation and the process (according to his individual interpretation). This is a kind of placebo effect. Even a situation as extreme as masochism or sadism can empower that sensation (again, according to each individual's interpretation). A placebo can be anything that the individual strives to achieve through a transcendent, supportive source, by carrying him out of the "quicksand swamp" and improving the way he feels. This conduct is engaged by our awareness. For instance: a mother who improves her son's negative emotions as he weeps operates by diverse and individual strategies.

A man who feels nauseous would feel relieved the minute he smells a lemon. **It won't be because the lemon carries**

in its scent various ingredients that cause relief, but rather it is an individual change in his sensation caused by the new atmosphere. The pleasant smell steers us into different and more pleasant places.

I will give you another example. Some believers conduct the revolting practice of pressing a pigeon's behind against a hepatitis patient's navel. Then, the pigeon's neck is pushed back until it stops breathing. According to these believers, the pigeon sucks the hepatitis out of the patient, out of his navel, and through its own anus. Supposedly, the pigeon dies of the hepatitis rather than the asphyxiation (what an outrageous distortion).

This patient truly believes in the procedure. The sickly sense of his "self" changes into a healthy one, clean of hepatitis – according to his new individual interpretation of this situation. And often it works: the patient's condition improves and he is on the road to recovery. That is a type of placebo.

The placebo effect doesn't work on a baby since it doesn't yet understand the symbolic interpretation of it (ups, downs, and overcoming are an essential part of any human being's life, with no connection to the taken potion).

The placebo pill only affects someone to such a degree as the individual is convinced that this magnificent pill is his life jacket. The "self" feels relieved by the support and so the body functions accordingly.

Therefore, a person who administers the placebo should be absolutely convinced of its efficacy. Among animals, for instance, the placebo is irrelevant. This is because without verbal understanding, one cannot convince an animal of its benefits and aims. However, theoretically speaking, if an animal could understand the intention of the placebo, it could affect it. That's why we reward animals with tasty food, and don't necessarily focus on the amount or size, to improve their sensation according to the "quicksand swamp" (in addition to the ego-enhancing treatment). The sensations of animals, much like ours, are affected by psychological factors. For instance, if we turn off the light at night on domesticated animals, they will feel relief from the heat. However, on a cold winter night they will feel better after turning the light on, as they will feel warmer. The light affects them for the better because it simulates heat (the light itself doesn't really heat). The sensation changes according to a new and individual interpretation to a specific placebo. This situation causes a tiny spark of relief in the situation of the "self" (like support). It is striving for a life jacket that would save it from the "quicksand swamp," thus causing a positive chain of reactions.

What is a chain reaction (regarding placebo)?
A positive initial spark.
A positive reaction in accordance with the initial positive spark.
Another spark, which is slightly bigger, since the initial spark proved its effectiveness.
A bigger positive reaction to the second spark. The belief in the placebo's efficacy grows bigger and bigger.
Another spark, bigger than the previous one, as a result of the growing belief in the placebo's efficacy.
An even bigger positive reaction thanks to the third spark, and the proof of its effectiveness.

And so on. A change in the sensation leads to a string of changes according to individual interpretation.

A patient who is treated with a placebo will feel relieved. Accordingly, he will find the strength to begin his new life. That is a life jacket of sorts. That person reinvents himself and embarks on a vibrant life – beating his illness. There is a borderline of difference between a diseased and a healthy environment.

There is also the effect of the placebo's effect. The initial effect relates to the individual, who, in any case, was on his way to overcome the individual phenomenon (like many others). The second and complementary effect that tipped the scales was the effect mentioned above. The more a man submits his body to an exterior substance in order to do the work for him, the more he abandons his resistance, his domination, and his shaping of the "self", all of which dwell within him. Accordingly, his body will deteriorate. Further degeneration of his body and mind will soon follow.

In light of this, we should go on and climb upwards to shape our "self." We do this by changing our negative interpretations of different situations into positive and constructive interpretations (as much as possible). In addition, we should also use the resentment instinct (See "**Resentment**" chapter).

Different methods and alternative treatments "work" by the same principle. They might help when things improve according to our personal interpretation. Therefore, the caretaker should be authoritative, and explain the healing methods of the pill or remedy given to the patient convincingly.

Be aware of healers who use witchcraft and other twisted methods.

I should point out that the placebo can cause a negative reaction due to a negative interpretation. I will tell you a true story as an example: This is a story of a child who cursed a highly important religious figure. His strict parents and the entire community admonished him for what they considered to be severely inappropriate behavior, so much so that he was convinced he would suffer a curse and grave illness. The young boy, as expected, who consequentially adopted a traumatic sensation as his new interpretation, suffered a heart attack and was bed-ridden thereafter.

In conclusion, the placebo's role is to change the distressed sensation of the "self" to a sensation of relief and elevation, by the positive individual interpretation of the "*Ultimate I am my own maker.*"

This is indeed a war for survival, but if you study its principles, "holding" on to the wheel of life would be less frustrating, and even challenging.

Resistance

What is resistance?

When a man pushes another man, the man pushed would instinctively resist his aggressor. If he didn't, he would fall.

When a man firmly squeezes the hand of another man, the latter resists and shrinks his fist so that the hand doesn't get crushed.

When a man gets punched in the gut, he should resist and clench his stomach muscles. There's no other entity (another mind or body) to do it for him.

When a man carries a sack of potatoes on his back, he must resist the burden. If he doesn't, his back will break.

When the stomach grows into a potbelly, instead of encouraging the "growing" phenomenon, one should resist it by clenching and pulling the abdominal muscles in an instinctive and precise fashion. There is no other entity to control our stomachs other than the "*I am my own maker.*"

This principle also applies to resisting illnesses. **For instance, when an illness breaks out, the patient must "resist" by pushing it away in an imaginary-sensory way, therefore, diverting the illness from becoming the "aggressor," and directing it to becoming the "aggressed," causing it to "flee with its tail between** its legs." This aggression, triggered by the imagination, operates via the image of "a supposed excess of lethal acid in my body that had slain every intruder and, in fact, had caused the disease's regression."

Regarding all you heart patients: you should use your sensory imagination to feel your heart; – a flexible, strong, muscular and efficient heart. If you feel preliminary stress that will inevitably be followed by cardiac arrest, you should not remain passive and "flow" with the situation. You should rather immediately and instinctively reach out and regenerate your heart, using a diverting, sensory imagination. That means suction, compression, suction, compression – operating like a quality (cardiac) pump.

Acts of resistance occur instinctively among animals, having no other choice. These are naturally acquired rules of survival. Nature doesn't carry a first, second, or third aid. The laws of nature thus dictate: "Live by the given conditions, resist, and fit in. Otherwise you get hurt, and when push comes to shove, you may very well become the main meal."

Unlike animals, humans have forsaken the laws of nature. We should be aware of that.

BACTERIA AND VIRUSES

As is widely known, bacteria are everywhere. No matter how thoroughly we wash our hands, even with alcohol, the simplest acts such as turning off the faucet could easily pass on germs. The mindless germs mock us. We have resistance systems, but for better or worse, there is no entity that will operate them for us. The resistance is carried out by the individual "self" and it does so according to the senses of resistance, resulting in such features as ruggedness, or passivity, or an array of other possible mental states (further explanations will be provided).

The massive use of cleaning solutions that pollute our world must be stopped. We have thus far tended to believe that a cold is created when a living organism catches a virus according to the "all or nothing" rule, meaning – either the one who catches it or the one who doesn't.
But that isn't so. Viruses are everywhere. As the "self" (meaning, every living organism) senses a regression in life, its biological resistance drops. The cold takes its place, turning from a minor to a mild condition. It depends on how sick the patient becomes. The remains of the illness depend on sensation. There are endless variations of the common cold. There's liquid phlegm and there's dry phlegm; a clogged right nostril or a clogged left

nostril, or both; sore throat; different kinds of coughs; different kinds of pneumonia; sinusitis and more. Their severity varies. Sometimes the disease is followed by a downhill collapse and sometimes by a triumphant, uphill recovery. This endless occurrence is determined by the "quicksand swamp." The living creature's sensations are not monotonous throughout its life. At times of collapse and passivity, doors open to viruses or other phenomena.

Viruses don't have predetermined dates for attacking or regressing. If that was the case, we would then ask: if the person infected with the virus managed to recover from it, why then did it attack him in the first place? Why him and not a different individual? Why today? In what temperature does the illness strike? (Some people live in temperatures that reach minus 50 degrees Celsius and don't get sick (the residents of Oymyakon, Siberia, to name one example). When passivity and a subsequent collapse occur, it opens the door for various types of viruses.

A sudden change in temperature could cause a cold only when the "self" that merged with the heat is not mentally prepared to merge again with the cold.

Sneezing isn't caused by a virus that set out to attack us. It's rather a reaction that is caused by a specific sense of "sinking." Sneezing is a non-voluntary reaction to that sensation.

A passive individual might be convinced that he is overcome by a sensation telling him "this will happen to me again and again." That is the determining "self," for better or worse, and it indeed happens. It can happen in the form of sneezing, hiccups, Crohn's disease, etc.

Apart from the physical aspect, our endless struggles are often manifested by our language, with words and with curse words. These urges manifest themselves among animals through the physical act of combat or (instinctive) body language, like stiffening their fur, growling, exposing their teeth and so on. The instinct that I chose to name the "I more" instinct, has caused in many cases the extinction of such animals as the saber-tooth tiger. The tiger's fangs developed through evolution to such a large size that it could no longer hunt and thus could no longer survive.

It is not inevitable that the individual interpretation of word puns affects our bodies. An individual who is hurt by an offensive remark will feel a sense of burden and inhibition in their body, and an instinctive reaction will lead them to create a counter offense. If they withhold the offensive behavior, the hurt that they feel will remain locked in the body of their *"I am my own maker"* persona, which might cause the illness to surface.

If the person who is being attacked is not hurt by the attacker's remarks, the tables will turn. It would be as if the poisoned arrow had flipped its course back to the attacker. That is an individual interpretation.

Remember – our urges to triumph help us improve ourselves.

In addition, be aware of the fact that it is only a personal, interpretation-based sensation that can be altered. Like: the winner is happy as a result of his victory, therefore I (as a spectator – not as a contender) am happy for him

For example: a man enters a contest. If he was convinced he would reach the first place, but disappointingly only reaches second place, he would evidently feel beaten, defeated, and humiliated. But if he had from the start aimed at the third place, and surprisingly won the second place, he would feel victorious. (These rules are unwritten in the brain, and have nothing to do with evolution or survival). It is the "quicksand swamp."

SLEEP AND DREAMS

How do we dream?

The body and the mind have their owners – the "individual self." We can compare the mental state of the "self" to the "quicksand swamp." Therefore, when the "self" is exhausted, the brain's efficiency level drops. By that it causes the body's efficiency level to drop as well. That is the "individual quicksand swamp."

Unlike common claims, **man does not engage in sleeping in order to feed and re-charge his brain!**

Life is mentally tiring and physically exhausting.
If life pushes the individual forward, he is less exhausted.
If the individual is exhausted and troubled, then he cannot sleep.
If the individual's life is very strenuous, he will sleep more and more in a chronically fatigued manner.
In fact, sleep means pulling yourself apart temporarily from the reality that surrounds you. I repeat, despite all scientific claims that the brain lowers its waves of activity levels and tells us to sleep in order to recharge it, I believe it is incorrect! **It is the "self" that is tired. The brain**

operates less well and acts according to the fatigued, tired "self."

Take a poor, exhausted, and worn-out man, who has a deep urge to sleep. Suddenly, he receives a phone call that changes his life for the better in a very significant way, and it takes him to the top! At that moment the feeling of exhaustion is completely gone and is replaced by excitement and new-found energy. Suddenly, there is no more need to recharge the brain. Therefore, the deep urge to sleep also depends on the subjective sensations of the "self."

The sleep of a man who is in a hostile and bothersome environment is rather light. But it is complete once the man is in a state of comfort or over exhaustion or both. Once the man overcomes some barriers and distances himself from his surrounding reality by taking a temporary break, he starts to have a mixed array of memories in his dreams, past, present, future; and also uses his imagination, his fears, hallucinations, and success as he links them all up. For good or bad, the type of sleeping and dreaming depends on each person's mental state, individually. We should point out again that **the brain is not the dreamer. It is rather the infinite "self" that dreams using the brain** (which he created during the evolution process, in the "no choice" process).

In conclusion, the brain does not dream. It is the "self" dreaming through the brain. The brain, a non-conscious, biological computer, doesn't dream for us to solve the riddles of our lives. Dreaming does not occur by will, since there is no entity to ask or to order specific dreams from.

We should immediately stop making fixed and redundant theories, which are senseless and inherently wrong.

SLEEPWALKING

Animals don't have the option to choose how long they sleep or how deeply they sleep, nor do they have the option to behave lazily. But for human beings, the situation is the complete opposite. The more a man is unable to hold onto life, the more it slips away from him as he collapses and slides out of the reality into sleep, which poses as an alternative reality and does not require any effort.

The sleepwalker carries hidden passions. The passions find the means to manifest themselves in his imagination and remain dormant, without **any awareness**, of course. The passions find their way out of reality. This is a legitimate and instinctive bonus as far as the sleepwalker is concerned. It is not something to be taken lightly because it is a dangerous situation.

One can escape from reality in numerous, dangerous ways like beliefs, religions, addictions (drugs, alcohol, gambling), violence, madness, and overeating. Sleepwalking is another extreme way of escaping reality.

Again, we are speaking of a man who could not "drive" his "self." "keeping his 'steering wheel' at the center of the road." "Keeping the steering wheel in the center of the road" expresses a conscious and disappointing reality. The sleepwalker, in this case, steers his "steering wheel" away from the consciousness and onto the subconscious, which lacks any responsibility, debts, or liabilities. He dives into another world, free of prohibitions, rules, norms, debts or survival rules. It is a getaway from lacking any freedom and an escape to an alternative freedom. Dozens of tons of serotonin and various other substances couldn't resolve this phenomenon.

Coping with Sleepwalking

Here is a recommendation of how to handle the sleepwalking phenomenon and several other ailments of the passive, modern man (like obsessive eating, diabetes, snoring, physical or mental weaknesses, etc).. We should get people who suffer from these ailments on a plane, attach a parachute to them, arrive at the heart of a jungle, give them a little push and wish them good luck. I am joking of course, but consider this – in the jungle you don't have a refrigerator filled with supplies; no comfortable bed in it either; no room of your own. So you cannot choose when you sleep, or if there will be appropriate sleeping conditions at all. It is a "no choice." We should embark on real survival in nature to "point out" the rules of nature as a term to control the "steering wheel" of life. Also, we should think more highly of ourselves. Turn ourselves from inferiors to proud rulers, the dictators of rules – and the ultimate survivors.

Again, I repeat: we shouldn't blame or pick on our brain. We shouldn't look for any magic solutions of various potions.

Have you ever seen a sleepwalking animal in nature? The answer is rather obvious. Nature's reality doesn't enable an escape from reality.

In nature, no other animal pushes another one to sleep or eat. There is a constant struggle and permanent threats to survival. In order to survive nature you must **survive**, with all that implies.

THE DIGESTIVE SYSTEM

The damned digestive system

The digestive system doesn't operate automatically. It is legitimately run by the "*Ultimate I am my own maker*" which is infinite. Everything that we have known of the way the digestive system works is no longer relevant. **The digestive system was made according to infinite sensations of the "self" as it is struggling and yearning for energy and life.** The instinct of longing for life and energy is the initial sensation at the beginning of evolution. It was formed, during the evolution process, into internal and infinite sensations of the "self" that are not a part of this aim of using just energy. The digestive system was made, and indeed is designed and conducted, not as it should, not as we want it to be, not independently, not logically and unnecessarily.

If by any force, an optimal digestive system could have been created, it would have had the following features:
- There would be very short intestines.
- The food would have been processed into dry droppings at the exit diameter.
- The food would leave the body after four to six hours.
- Food would come out without any effort or disturbance,

with absolute control over the passing out of food, and with an enormous burst of leading energy.

In short, the result should be: **digested food, no less!**

But since the digestive system doesn't have any other supervisor, **it is controlled by the "self" – with all that implies.** In fact, the food stays in the body for a long and unnecessary time. It sometimes stays for too much time (among individuals who lack energy and a firm grip on life). That is a sticky sensation.

The body doesn't differentiate the past from the present and the future. Because of this, when a living creature eats, the food is absorbed along with all the toxins and surplus vitamins. Only then, does the body drain itself from the toxins and surpluses, according to the infinite state of the individual "self." There is no other entity other than the "self." The digestive system is created and is operated according to the present, according to the interpretations of the situations that the man had felt in the past, present, and in his sensations of the upcoming future.

Every given moment in the process of evolution was a present moment, which consequentially created the current shape of the digestive system, without any reference to the past or the future. This shall also be carried out later throughout the evolutionary process. If the mind had known of the future, then all this sloppiness wouldn't have been our share. On the contrary, the body traces toxins and rejects them before they are absorbed in the system. This is not a "clever" operation, to say the least.

There is no maker of the digestive system. If it had a maker, he would have designed it so that it would only process what the body needs rather than absorbing everything – toxins included. Unlike a car maker, who plans the vehicle in advance, from start to finish, and only then sets out to build it, our body is built in stages, according to the present situation. In other words, **our body is built by what is right for it at that particular moment, according to the subjective sensations of the subconscious "self."** All this is done regardless of the future. There is no mechanism that draws conclusions, and learns from the past or prepares for future results. It is only the "*I am my own maker*" that created the digestive system, with all that is implied.

The initial instinct is the sense of yearning for energy and life. Processing and draining food, on the other hand, are later stages that comply with rhythm and leading sensations.

If a man drinks one drop of poison, the brain doesn't necessarily recognize it as such. The toxin is absorbed in the body, and only when the "self" feels that it is poisoned does the body cleanse itself, according to the individual poisoned sensation of the "self."

Burdens that manifest in sentences like: "Why did this happen to me?"; "On the one hand it's right for me, but on the other hand it's wrong and difficult."; "I have no other choice."; "I had to."; "Against my will."; "I wish it weren't so."; "Why couldn't I have done it differently?"; "Why me?"; and, "Grief," "Disappointment, and" "Lack of motivation." All these sayings represent hardships that don't dissolve into thin air (the same goes for orgasms). These are subconscious hardships that a man, or any living creature for that matter, feels in his gut; **and they are expressed (throughout evolution) in a redundant, prolonged, and disruptive activity of the intestines.**

The digestive system is operated and acts only by the infinite sensations of the "self," in accordance with the endless various styles of the "quicksand swamp." Once the individual is at his "rock bottom," his digestive system, like many other organs in the body, reacts with fibrillations, constipation, ulcers and gas.

Since the digestive system is highly sensitive to our mental state and therefore acts by our conscious and subconscious sensations, each living creature must feed upon its passions to have a satisfying conduction. Instead of a person focusing on eating "healthy foods" to stay well, he must feed (conscious and subconscious) urges that manifest in his gut, among others. Vitamins or good bacteria cannot heal a damaged digestive system. Instead, padding the "self," fulfilling oneself, and carrying out hidden passions (including subconscious passions) will miraculously improve the digestive system. So will psychological therapy.

In addition, throughout evolution, man has suffered many humiliations. Whether it is a king, a tyrant, a domineering husband, a god or any other frightening, domineering persona – whether it is a helpless man – the difficulties will bottle up inside, in the gut. These emotional loads cause bloating, distortions, and many disruptions in the digestive system, causing problems in the proper conduct and elongation of the intestines.

Thus we conclude that behavior inflicted by the will to "overcome", that is, to win, to fight, to curse, to insult, to drive fast and more, are not more than "elbow shoves", which we instinctively use to protest, climb, and get "on top." The more superior and uninhibited our sensation is, the better will be the conduct of the digestive system, in direct relation to the freed sensation.

Food and digestive disorders

Unhealthy food?! What about a breast-fed baby who suffers from digestive disorder? Why, in order to process milk there's no need for the intestine's sloppy length. The baby's "self" produces (other than earaches and others) trapped gas by its frustrated, withholding sensation... (Bacteria?).

You should, for instance, walk the baby in a stroller with jagged wheels, where it can see different views. This bumpy ride will expand its sense of freedom and improve the functioning of its bowels.

Each person is a unique, infinite individual. He senses and sees the world through his own eyes. Therefore each person feeds upon different kinds of food, by his personal tastes. The same goes for animals.

Let's conduct an experiment. Ask a volunteer who is mentally suffering on a daily basis (like a beaten and defeated obsessive gambler) to eat a nutritious dough that contains all the so-called healthy foods. Preferably, the food shouldn't be to the liking of the volunteer. No change will occur in the quality either of the digestive system or in the physical health. The result might be the opposite of what's expected, and what is normally thought to be, due to surpluses, poisoning, aversion and feeling sick from the food (in an individual manner). Thereby the food will be processed according to the individual's interpretation and sensation in regards to the rejected food inside the body.

The body needs a small sum of the energy that is in the food a person eats. In fact, each living creature eats not just to produce energy from a biological source to the body. The practice of eating is a part of life. It occupies you, relieves boredom, fills up empty time, and releases frustrations. In short, it fills up our life with "delicious" snacks. The energy is out there, but it doesn't fully manifest itself, according to the "quicksand swamp" (like chronic fatigue and exhaustion). For example: the sugar level in a diabetic patient's body can rise up to seven times the norm, despite the fact that the food he eats is normal and well balanced.

Candy and sweetened sodas sweeten our lives, by elevating our sensations. That is also the effect of adding salt to your food (changing the sensation according to one's taste).

The rhythm of conduction and bowel movements

The rhythm of conduction is not continuous (besides the conduction energy) and it is accompanied by energy that is restricted all throughout the intestines, along with the processing style. It is not continuous because it is set by a variety of individual moods of the "self" like recessive feelings, various blocks and lack of passions and plans, dissatisfactions, living with memories of unresolved issues, a forced present and a bleak future, sensations of various "short circuit," and therefore infinite types of conflicts. Or it could be the opposite: lust, happiness, and positive energy, with an elevated sensation and a good sense of conduct. And maybe it could be a little of both.

May I remind you that the digestive system operates according to its owner's personality as it is affected by such traits as optimism or pessimism, along with various other personal characterizations. These sensations are endless. That's why the term "infinite" plays a role in all our lives. For instance, if you win a monetary reward or make it to the top, the brain still doesn't operate by a mechanism possessing such conditions as: "If you win, I will make your body work more efficiently." These are all individual sensations, triggered by the interpretations of the infinite "self." If the goal is important and desired by the individual, and he wins, he will feel relieved from feeling stuck. This feeling will transform into a freed sense of conduction, which would also manifest itself through the "*I am my own maker*" digestive system. This is individual interpretation.

In comparison, the winner *will* most likely experience **subconscious** internal affairs like mixed feelings, shock, and confusion and he would be overwhelmed by conscious and subconscious ponderings. "Maybe money is the formula for happiness?" Or "Maybe winning the money will lead to negative reactions by greedy people or exploiting friends?" and so on.

Blaming the body (the digestive system) is irrelevant. We must, therefore, pad the "self", just as a good mother encourages her frustrated son and improves his emotions.

You should make plans to succeed in the present and in the future.

Forgive the mistakes that were made in the past.

Learn your lesson.

As far as our bodies are concerned (and particularly the digestive system), its owner's future is concrete and actual, as it is determined by its focus and its importance. And it is also true regarding the past: if a man goes to work overwhelmed by thoughts of his agonizing, demanding, and humiliating boss, or of disliking his job, he will feel what will be coming even before showing up at work. He might definitely experience these feelings as disruptions of the digestive system or in other physical disturbances. Coming home from work, he will think about the obvious resentment of his bitter wife and his feelings will again manifest physically, causing a disruption to the digestive system.

Meaning, the present sensations dictate the operation of the body by these subconscious, individual interpretations.

Conduction and the draining of the bowels

The rhythm of conduction isn't ideal. It is irregular, not permanent and random. It is carried out by the infinite sensations of the "self." Sometimes the "self" is at its lowest and therefore lacks any passion, energy, interest and enthusiasm. This is caused when the "self" is reserved and blocked in everyday life. And at other times the "self" transcends and is enthusiastic, happy, and full of passion, lust, interested and excited. It loves and is loved, it is attracted and attractive in return, feels victorious – in other words, feels a sense of conduct through life. And sometimes, of course, the "self" carries a mixture of all.

Life obviously isn't made up of black and white situations. Rather, every living creature has varying moods, experiencing his own ups and downs. Digestive interruptions are inherent to the sensations. Because sensations lack permanent order, various changes, disruptions, interactions, and reactions occur within it, as well as the digestive process itself. Massive disruptions are an extreme result of emotional ties that are like a "short circuit": Like "I love her, but she doesn't love me." "I want her, but I can't have her." "I want, but I'm not allowed to." "She loves me and I her, but another person is disrupting the relations between us." "I didn't feel like it, but I had to." "I have no choice." Or a strong excitement, endless and subconscious conflicts, and so on. All this is carried in a temporary, asymmetrical fashion.

An extreme lack of conduction, such as constipation, is caused by a sticky sensation of the "self" being stuck, or worse – a sense of recession. In simpler terms: an inhibition caused by unfulfilled desires, while life is slapping him in the face. A negative reaction, just like any other phenomenon, can stay in the digestive system for a day, two days, one week, two weeks, one month, two months, a year, two years or forever. This depends on the situation the individual is in, along with his mental state.

An image that demonstrates the subjectivity of the digestive system according to the interpretation made by the "self" is a financial situation. A person who gets a check without coverage or doesn't get the money he deserves, experiences disruptions or a sense of burden in his heart or throat, ulcers and so on. A man with a billion dollars in his bank account feels happy and "on a roll," only until he discovers that his competitor has two billion dollars in his bank account. Then he is likely to experience similar disturbances according to his new interpretation. **It doesn't matter what the "self" actually does. What matters is how the "self" interprets that situation.**

Another example of the subjective interpretations of the "self" is boredom. There is voluntary boredom that is intended to disconnect or to rest. And there's an imposed boredom that is accompanied by the following thoughts: "I want friends, but I don't have any." "I want to play, but nobody wants to play with me." "I have friends and I play with them, but they keep beating me." The person whose boredom is imposed upon him will have thoughts that create a sense of burden, weakness, and fatigue. It also manifests itself through such physical symptoms as sickliness, thin and sensitive skin, sensitivity to the sun, general sensitivity (including noise), disruptions in the digestive system and a lack of conduction.

After setting up part of the principles, we now realize that this is not the way in which the digestive system needs to operate.

The biological procedures in our bodies do not operate like a machine. Every living creature survives **mentally**. This sense of survival is infinite and resembles the "quicksand swamp." The digestive system, along with many other organs, operates by the subconscious "self." We consciously eat and swallow our food, but this action is motivated by the individual subconscious. Meaning, a moment after we swallow, the consciousness is no longer relevant. Since the dawn of evolution, man has been mentally "sunk," full of mental hardships and helpless. These situations led to a lack of conduction. **This lack of conduction made our intestines grow,** because there was no mediating entity to maintain it at a normal size. The difficult sensations manifest through the inside of the gut (the digestive system), and therefore digestive interruptions, such as mal-conduction, have been one of the physical burdens that humanity has had to suffer throughout the evolutionary process. All that had taken place throughout evolution is sealed in the genes, for better or worse. The result: long, sloppy, and burdensome intestines.

Therefore, we should "live" in all aspects of the term: fulfill our cravings, feed upon our urges, and be happy with what we have.

But on the other hand, and most importantly, we should change our interpretation of situations and events as much as possible. In addition, each person should educate himself (his "self") that "I'm happy this way and I have no over-the-top demands on life."

Sometimes, a chronic lack of conduction causes distress and lack of energy, and as a result can also cause weakness and passivity in regards to our bodies. This can indirectly lead to obesity.

Sensitivity to certain foods is individual for every person; to his own personal taste, which is affected by prior situations. Like when a certain food reminds someone of a tragic or unpleasant event from the past. The same goes for the present and the future, when individuals are convinced that the food that is within them is repulsive they will inevitably be repulsed.

When a man drinks water after eating watermelon or grapes, he feels that the ingredients don't go together well and will develop gastroenteritis. It's not because the watermelon and water don't actually go together, but rather because the "self" has an internal sensation that they don't. The digestive system doesn't independently process the food that we eat, but rather the "self" does. If he eats vegetable soup, his sensation will become better.

It sometimes happens that the body exhibits extreme negative manifestations as a reaction to some situations: when a man has a bit of what he considers to be repulsive food, his sense of repulsion normally causes him to **feel** a minor, concrete gastroenteritis (which is not just in his sensations). Suddenly, his fears have come true and therefore, his gastroenteritis gets worse. Then he is convinced that he had rotten food and his situation keeps getting worse and ends up as an acute gastroenteritis. All this happened despite the fact that other people ate the same thing and did not suffer any side effects.

A different possibility is that all the people who had eaten this food had contracted the gastroenteritis due to a chain of reactions (See "**Copying Sensations**" episode). The chronic diarrhea phenomenon, just like other phenomena characterizing modern man, exists among individuals who have been "tamed" by their parents to be ego stricken, delicate, and passive, and to lack resistance and be sickly. This weakens the intestines (among other side effects). An individual who is convinced and expecting this frustrating phenomenon will have a sensation that eventually comes true over and over again, by self intention, for better or worse.

The intestines have plenty of "conduction energy," hidden somewhere and manifesting itself. It doesn't clean out the bowels by a conscious decision, hence the will itself cannot determine that schedule and rhythm of the conduct. If the birth labor sensation was pre-made or voluntarily administered, you could have cleaned out your bowels all at once.

Theoretically, if the body were to operate and function by will, and if the brain had any understanding, the brain would have given an order to the intestines to dominate the emptying process instantly, or perform it in any other desired manner.

Some remedies cause certain side effects and don't actually cause relief. But it ought to be stated that the sensations are unstable. Therefore, the previous remedy is not necessarily right for the current situation.

In conclusion, the conduction process (and its sensations) of daily life operates a biological conduction, among others, in the digestive system.

Ulcers

Ulcers are the situation in which the "self" is at a point of "eating itself" from the inside, according to its self-destruction sensation, caused by its individual interpretation. These sensations, that the individual builds up inside rather than pushing outside, are relived by him in the present, or he experienced and lived them in the past, or even felt the upcoming future experiences or any other combination.

What causes this is usually a different individual or others, who control his sensations, causing him to feel bent over and forced. In other words, the "self" builds up strong passions inside of it that do not manifest since they are confronted by an external thwarter when it comes to shutting people up, oppressing self-expression, holding back passions and so on. Feelings of helplessness surface, as a "short circuit," thus causing the phenomenon of "eating oneself up."

There is no direct relation between ulcers and food. When a man or an animal is hungry, he thinks "food" and his body operates by his consuming sensation, secreting acid according to his hungry sensation. If food doesn't come in at that moment, the stomach will eat itself up. This situation might cause ulcers for those who are sensitive to ulcers, due to factors I mentioned earlier (those who drink alcohol and eat spicy foods regulate the acid level in the stomach without being aware of them doing so).

When there's a craving for a **change** in an intolerable situation, the "self" eats itself up, like when the situation doesn't match the craving (in a kind of "short circuit"). This resembles the situation of a man stuck in traffic, though he's in a great hurry. When it happens, it seems as though the car won't accelerate "out of spite", as though it "insists" on driving slowly. The craving is meant for us to get to our destination already. Traffic collides with his craving. As a result, he will eat himself up.

Ulcers don't usually exist in dog races, since no rider will dictate their moves and push them around. Race dogs enjoy freedom of expression in regards to their abilities; therefore, they run with enjoyment, love, and an urge to win. That is well-balanced conduct. On the other hand, when a dog is tied in chains, alone, under the skies in the hot summer midday, dropping his bowl of warm and steaming water with his leash, the same leash that is wrapped around the tying base and shortens into zero distance. He only gets food occasionally – when his stupid, mind-numb owner leaves it his leftovers. The dog is bugged by biting, flying parasites. He is bored, humiliated, and is jealous of his free-roaming friends as he cries and wails to get free of this scrutinizing agony, while his owners use a remote-controlled belt providing an electric shock to shut him up. In addition to all this, the child amuses himself with the remote control for his own pleasure. No one addresses this hair-raising situation and the dog's agony. He will suffer from ulcers...**among other ailments**. This is the dark, human maze.

We should tie these "human creatures" with the inflamed yet big brains, under the same conditions so that they will develop an understanding of the other.

Unlike race dogs, race horses run freely in an imposed fashion by the whip, not by their urges and desires. Therefore, they might also suffer from ulcers, according to the agonizing sensations. The next forced execution (race) and the ones to follow will be accompanied by a sense of anxiety.

Treating Ulcers
You should step out of this situation determinedly and

immediately. A person in this situation should rise above his current self. Be someone else: an active figure who will powerfully push him until he elevates "beyond". Like a good parent protects his child religiously, the conscious and understanding "self "must protect the non-understanding, unaware, and unconfident "self." When a man's feelings are hurt, he should adapt his attitude and extreme interpretation that the non-understanding "self" relates to different situations and words that he heard. That's because each individual has the victory instinct. Responses like sarcasm, an insulting comeback, dominating, mocking, etc., are legitimate weapons.

You should be aware that ulcer sensitivity indicates a degenerated mental state and a personality subdued to society and to various situations. Eating and burning internally reflect the "self" struggling to change the situation when the will doesn't come through. The body doesn't operate by will because there is no entity to listen and answer to our needs. Once a man is aware of these laws of nature, paths that had been dark to him up until then now brighten up, as he finds new ways of changing his character and expressing himself without inhibitions until a free conduction occurs.

Among Animals

Among animals the same principle applies. Creatures in nature have free choice. They can choose whether to run for their lives, or abandon the herd and join the others. They can react as they wish or reconnect with better, more comfortable friends of their choice.

Animals are not race carts. They, much like us, have preferences; personal tastes; characters; passions; urges; lack of urges; jealousy; preferring the neighbor's back yard, preferring a different caretaker, preferring different company; having fears, loathing, despair, boredom, lack of passion, lack of appetite due to some event or thoughts; yawning; loss of sight; stress, past traumas that lead to future traumas; paranoia; mental illnesses; chronic weakness and so on.

An animal is a social being that lives in a herd outdoors. When it runs, it encourages and provokes an urge in the other, in an energetic enthusiasm. If the animal manages to escape from its predator, it shows it is victorious, and is proud in its noble body language; its tail and head are held high, and it is strong and full of energy; its metabolism is usually high, and it experiences synergy and a rise in the level of good substances in the body as the toxins and harmful substances decrease.

This is the "scales principle" of feeling victorious and of proper conduct. We should imitate these terms and apply them to domesticated animals. Meaning, we should train a horse with at least one other horse, so it won't carry itself around aimlessly, and we should also do it in order to improve its mood and urges. It should be bribed with love, compensation, and a sense of belonging. We must treat a horse with respect, and never degrade and humiliate one.

I recommend training horses and humans once every other day. One day to accumulate mental energy and another to unwind it. I say this so as not to create a routine inflicted recession. Humans and animals are not machines.

HORSES

An important message: the aim of this book is not to teach anything about horses. I use horses as an example because the principles are completely the same.

Usually an animal such as a horse is forced into a stable and the man is the one determining his fate and burdening him with tasks that the horse doesn't want or like, or rather he lacks the passion and the urges to fulfill them.

In this situation, the horse's "self "eats itself up" from within, without a possible reaction, in a "short circuit" of situations that are interpreted as "eating yourself up." The horse's condition deteriorates under the burden of the man's expectations of it. From here on, a chain reaction is carried out, leading to a mental degeneration and utter collapse. Suffering ulcers in this situation is unlike any other pain in the world. It burns and eats up the live flesh by consuming acids that the "self" creates, depending on the consuming sensation from within. The more he focuses on the scorching pain, the more it passes on more acids. In this situation, the energy level drops bottom low.

In this book I offer a way of using the brain correctly, by changing our attitude and world view, rather than managing our lives by fixed norms. There is no alternative for curing ulcers other than a mental treatment. We should recognize the sensation of others and learn to read them. By doing so, we will learn to recognize the feelings of the horses we tend to. For instance: jealousy could eat at the horse's soul and humiliate the horse, just as it does us. So if we attempt to mate a mare (which the **horse considers to be "his"**) in the presence of another horse that is locked up, the other horse will be devoured by the sensation that its spouse is taken from it to be with another male, and it will therefore be forced to deal with this sensation in a way that would devastate it, as it would any *human* male. Another benefit of understanding the sensations of other living creatures is that we could avoid making such a mistake as feeding some of the horses with a tasty mixture while neglecting one of them; for we would understand how consuming its feelings of jealousy could be.

The invasive procedures operated to diagnose ulcers are irrelevant because the burning sensation isn't stable enough and may alter, for better or worse, as a result of the focus on the consuming causes. Along with focusing on the causes, the emphasis on the frustrating, burning sensation itself only empowers it. Moreover, the sensation always overcomes the remedy that is administered. That is the infinite "*Ultimate I am my own maker.*"

There is a common and fixed practice among horse trainers called "launch." During this practice, the trainer stands in the middle and rushes the horse using a whip, to make him run in circles. It's a demeaning, exhausting, consuming practice and it should be stopped immediately. (See **"Urge and lack of urge"** episode).

Laminitis Disease and the Fault of Mankind
Laminitis is a disease that attacks horses. It causes distortions and hurts their hoofs.

Illnesses such as this are caused by various factors. Some have something to do with inappropriate food or the surpluses of that food. If that is the case, then what is the connecting link between inappropriate food and Laminitis?

The digestive system doesn't operate for granted. It operates by the "*Ultimate I am my own maker*" of the horse. This behavior doesn't serve the will, because the "quicksand swamp" stands in its way. Inappropriate food only burdens this frustrating conduction further (see **"digestive system"** episode).

Inappropriate food, like: spicy foods, foods containing acidity, or compressed food burden the horse's system, cause a lack of conduction, and as a result lead to chronic gastro illnesses.
The black or white phenomenon exists among "animal lovers": As they understand it, as long as the animal doesn't collapse, everything is all right. It's another distortion in the human's conduct.

As mentioned previously, the "*Ultimate I am my own maker*" operates and manages the body infinitely. It is the horse's "self" that is crying out for bowel conduction through the frustrated will that can no longer pull through (meaning, the will is not in charge);

thus causing weakness, distress, and lack of energy. The horse focuses on the pain and the digestive disabilities. He redirects his attention from the rest of the body (the legs) and focuses entirely on the pain in his frustrated guts. This manifests itself in the form of poor blood circulation and lack of oxygen. The legs, therefore, chronically lack energy, and this situation causes biological distortions and Laminitis (among others). As a result of this situation, it seems as though the legs are no longer owned by any-body (literally).

It's like a foot going numb in moments when a sensation appears as though one is sinking into disharmony, which matches a biological disharmony.

An extended recommendation for concentrated food manufacturers who manufacture food for horses in pellets:

Food must be **tasty** and have **volume.** For example: squished oatmeal seeds or barley, raw wheat bran, raw wheat hay (recommended, but not necessary), a salt powder, tasty oils (to "nourish" the sensation). It should be divided into **small** portions (one kilogram top), both because of the limited biological structure and to break a frustrating routine. Hay is still condensed into cubes and it doesn't suit horses because of its density. It's like eating gravel. Here is where the principle of the quicksand begins. This causes an extreme absence of conduction along the intestines, which causes a sensation of great distress, a lack of energy, and weakness. Ignorantly, they add compacted feed, to supposedly add energy, when, this actually causes severe physical phenomena, such as: tingling of the legs, pains, brittle bones, twisted hooves, and exposure to diseases. This leads to massive and completely irrelevant veterinary treatments, which leads to the owners' disappointment and frustration. Then the owners respond with even more aggression and violence, which causes the horse's collapse, criminal neglect, etc.

That was an example of a ticket to hell.

Alternatively, raw wheat hay, which is **gently** packed into cubes, should be made. As for the quantity of food, you shouldn't force into an animal's body "highly enriched" nutrients like protein, mixtures, and vitamins to accelerate their performances. This belief was born in the minds of imaginative people (explanations will follow). This moronic behavior causes a lot of side effects, distortions, and a general helplessness, which in turn causes chronic illnesses among horses ending with unnecessary deaths. Most times, people tend to act with respect to the "all or nothing" rule. Their initial enthusiasm pushes them to over-feed the distinguished animal with tons of redundant mixtures, food, and supplements, allegedly, to improve energy and performance. But when the enthusiasm dies down or when we become disappointed with the animal, we discard it like a damaged object that has expired. This phenomenon is a crime and criminals should be punished. There should be an international law allowing only licensed caretakers who know what they are doing (according to the parameters set by this book), and who own farms to keep animals and establish an authority in charge of the caretakers with unannounced visits and (positive) informing.

A recommendation for the ultimate mixture will be extended in the next book.

A different example: a depression caused by various reasons like parting from a friend or loneliness, or being treated in a violent manner. The depression is accompanied by a lack of energy. Wrongfully, we add proteins and mixture to the food in hopes of improvement. But his condition keeps getting worse. And when it comes to carrying out a task, in his inferior position, he collapses.

Another extreme example is shock. Recognize the well known trepidation. This trepidation equals its **sensations**, when it begins to resemble fear and sudden shock. Like when a man suddenly strikes him with the reigns and there's a bridle in his mouth so that the animal is totally controlled and cannot instinctively react. This sudden shock causes a deep and immediate mental blow. With a flick of an eye he detaches himself from the energy that leaves his body and legs that support it, leading to a meltdown. He carries the shock, trying to get up on his legs that are lacking energy indirectly, with his strong will and longing for life. The Lamina bone in the hoof turns (in a shortcut) in accordance with his alternative (different) condition: a situation typical of his present traumatic situation.

The cure for this illness is asking for the horse's forgiveness by body language – patting, paying attention, hugging, offering bonuses in the shape of carrots, a gentle and non-threatening touch. He should be kept company by an inferior friend, like a donkey or a goat. The best option is to cut him off from males horrifying presence, and move him on to the care of an empathetic female caretaker.

All the above options are a combination of various causes that were hidden up until now. But that's not all. In addition, there are many issues with the horse's legs that are caused by the poor quality of the hoofs (distortions, cracks, decays, mass dissolving and so on), in addition to other physical distortions. That is the "quicksand swamp." There are many other phenomena that are caused by unknown factors, as they have been known up to the publication of this book. All the factors mentioned above (including the "quicksand swamp" and" improper food) cause a lack of energy, weakness, and passivity. Those create, among other things, lack of coordination in body movements, causing each of the legs to harm and injure the other, which eventually results in massive, misled treatments by vets and horseshoers (see **"Urge and lack of urge"** episode).

As for routinely treating parasites and worms, they should be treated, but only when there are **actual** worms. This is to prevent the declining efficiency caused by overdosing, like taking too many antibiotics, which renders them ineffective. Man is an "expert" in making the same mistake twice.

The reason for the existence of worms is the "quicksand swamp." This means that a frustrating routine and a lack of content cause passivity and lack of "resistance." The lack of resistance causes the appearance of multiple worms. Multiple worms cause a discomforting sensation, and this sensation brings about a collapse by the "scales rule." The principle also applies to humans, but the "self," the conscious individual, with feelings of disgust, might catch it due to his awareness of the phenomenon occurring around him and carrying out the phenomenon in reality (see "**Carrying out illnesses**" episode).

All the moronic theories regarding the causes for the existence of the worms (catching them from a different source, the cycle of the worms' hatching, etc). are irrelevant.

Take the pheasant, for example. There is an incredibly beautiful pheasant living in the woods of India. This remarkable bird is not made for captivity. It constantly looks for a way out of its cage. But what can we do about the fact that a man treats it as a decorative object and nothing more? This constant sense of stress, crying out for freedom which it cannot fully fulfill, causes distortions in its feet. Therefore, it carries a distorted mutation for future generations. Does it make any sense to you, capturing birds or other animals in a **tiny** cage for the rest of their lives? Should we keep on doing this because it is already being done? Can the cause for the beautiful feathers becoming weathered and constantly shedding be depression, which in itself is a result of the lack of serotonin?

There is a bird that is locked for life inside a tiny cage, so that its chirping can be heard and enjoyed by its captors. When the bird has company, it doesn't chirp. Have we been so deaf to the fact that these chirps are in fact cries out for freedom!

WHALES

For years now, scientists have studied the enigma of why whales swim up to the shore just to take their own lives. Still, this question remains unanswered. But why *do* whales do that?

An old, sick, or exhausted whale needs to keep afloat and breathe, but how can it do this in its miserable condition? It swims up to shore to support its body upon the surface, and to get its ventilation hole above water level. What will happen if we were to bring up several of these whales to one beach? Let's take a mother whale in order to answer that question. A worn-out, sick matriarch. She swims up to shore, with her daughter accompanying her. Her daughter does so even if she doesn't breastfeed any longer, simply because her mother is her entire world and without her she is nothing. Theoretically, the young whale could survive alone, but since animals are not machines, they are attached and love just like human beings. That's why the young whale follows the mother as does the rest of the family, just as they did when the mother was healthy. It's a custom for a mother to call the shots. The baby chick, for example, cuddles underneath its mother's wings even when it's warm. Mother hen wraps her wings around her chicks, being protective of them. And it's not the hormones either. It's a sensation that says, "Life is miserable without belonging to a mother." It's important to mention that the family members that join the mother might not be aware of the possibility that they're approaching their demise. And when things hit rock bottom, they all get trapped on the beach and die. They're stranded in a state of confusion and shock due to the traumatic situation they are suffering from. In these sorts of situations, every living creature loses its senses. This confusing situation could cause a tragic chain reaction.

BODY FITNESS AT ALL COSTS?

An important rule that one must embrace, which applies to men as well as animals, is: **physical training isn't necessarily the equivalent of fitness and health when it lacks the love and urges to do it.** It's true that heavy breath and sweat will occur in response to training, but that doesn't mean that the person training will improve his fitness and build up his muscle mass. In this case, the physical manifestations that occur in the body are as follows: a sense of burning in the trachea, fatigue, yawning and retreating, and most likely a sore kidney on the side. All these symptoms begin accumulating after a while and turn into a chronic mental fatigue: an expression of a dominant side, which is contrary to the "no choice" side. The result: aching joints on one side, or generally one of the sides suffering from chronic pain.

A training session that is carried out without any passion can also cause bone ache and their microscopic dissolution (loss of mass and microscopic fractures), localized stiffness, concentrated pains, and the loss of passion for training.

A horse, for example, which trains alone with no urge or passion, and is trained by being pushed, won't necessarily be fit. Alternatively, we must lead it toward the horizon, nurturing the urge within it on the way home.

Let's carry out a ridiculous theoretical study: take a nude volunteer; spray him with a substance that attracts flies; and place him in an area swarming with flies.

Question: will the volunteer develop muscles and fitness after a massive attack by the flies?

The result will be as follows: a sense of despair and weakness, a sick stomach, a numb body, a weakness of the muscles and a general lacking in body mass. The opposite result to what would be expected.

Let's say we have given the volunteer all the supposedly efficient substances: protein, vitamins, and steroids. His condition will only worsen and become life-threatening. His weak body won't be able to take on the surpluses.

It's a funny example, yet it definitely provides food for thought.

Animals don't train in nature. When a predator launches an attack, the prey can't complain to its attacker, saying he didn't train and that he isn't in shape. The competitive and threatening reality pushes us to be in a constant sensation of fitness.

In conclusion, the body works and carries itself according to the sensations of its owners ("I am my own maker"). Physical fitness means that the "self" feels in shape at that given moment and is loaded with plenty of urge, passion, and energetic excitement, regardless of the training and the pumping up of energetic meals.

HUMANS AND DONKEYS

Thus far we have all believed that animals are made out of different matter than humans and, therefore, lack understanding of things; get used to them; forget certain things after they happen to them; or are made for a certain type of treatment. In short, that they are not living creatures like us. These are expressions devoid of any logic. Besides horses, the poorest creature is the donkey, since men treat the donkey as an actual work beast and nothing else. Man carries that message to his children and this twisted tradition passes on through generations. All this is accompanied by brutal abuse, even to this day in the "smart" modern age of mankind. Ironically, it is the donkey that, unlike its appearance, is characterized by its innocence. I would rather die than be a captive of men if I were the donkey. That is a non-existing option for the donkeys. The donkeys lose their appetite, among other things, when they lose their lust for life.

In addition to the miserable life a donkey leads with a

human being, it is mostly used for labor from an early age. With a collapsing back, the automatic response by a man is to whip it even further. Brought back from slavery, his legs and hands are shortly tied so he won't run away, under the open skies, in hot weather when it is not being routinely attended to. It only gets food and water sometimes, just to be kept alive, because as far as men are concerned, it is a donkey and it's made for it.

Sometimes kids get a donkey to play with, with the parents' consent. All day long they ride it in turns, forcing it by whipping it with a stick. They tie it with a short rope at night, with no food or water and without trying to keep the flies off it. The next day they continue playing with their "toy"; for if they don't, it would just be forsaken out there, starving, thirsty, exhausted and in agony until its demise.

Are they cruel children? No, that's the result of the messages they have been receiving from their mindless parents.

This conduct occurs in every sector, especially with parents who only know how to produce many children, thus perpetuating the cycle of poverty.

If there was some entity in charge, it would only have been

fair if it punished all those so called "men of law," who compromise and turn a blind eye, and by doing so actually cooperate with this conduct.

The internal sacred order calls every minded person to take action in the matter, because, being unaware of our deeds, we sin over and over again.

There are labor horses, especially in overcrowded and poor, multi-ethnic areas with high birth rates, which are going through hell. Their owners are ignorant and lack the minimal understanding of how to take care of horses. And so, they are doing it completely wrong. They cage them in improvised and dangerous stables. They also bridle them with improvised and inappropriate bridles which wound them daily as they work. In addition, they beat and lash them regularly with a whip. These horses usually suffer from anorexia (and other ailments) as a result of their traumatic lives.

There are many animals out there agonizing, screaming for freedom from the "smart" man, the human creature.
As for the claim that they are "made for it," keeping a horse in a fly-infested area could lead to an aversion to and intolerance of flies. There's a sense of a disturbing burden,

as frustrating as an itching sensation, leading to a chronic physical itching sensation.

Staying outside, on sad nights, subjected to the rain and freezing winds, the horses develop a sense of physical stiffness and lack of flexibility, and therefore, have a chronic sensitivity to the cold, numb, stiff body.
The fur doesn't protect them from the cold and the freezing wind when it's soaked with rain. Let us not look for any reason in evolution.

HEARTBURN

Heartburn is a reaction to various situations occurring in accordance with the "self" "feeling heartburn." This sensation is caused when yearning is bottled up, causing the "self" to sense that there is not even a gleam of a solution or idea in sight. This occurs when the individual "self" doesn't find itself; doesn't know where to begin; or when it suffers from a guilty conscience. These are the various types of regression sensation. In such situations, the "self" resembles the sensation of heartburn.

The body doesn't function automatically. It is only by the

command of its owners (that is the infinite "self") that the body manages to function. Each individual breathes differently. If the way he breathes is twisted, the breaths will be accompanied by toxic and burning oxygen. This sort of personality is very sensitive to acidic foods and alcoholic beverages.

In this case we had better change the way and the manner of our breathing until we find the optimal style. We should keep the specific style of breath fixed in our memory and focus on it.

VOMITING

In order for conduction to occur in the digestive system, one must be in a state of high, with a sensation of flowing and contractions, upon which the bowel movement unfolds. Vomiting is a condition of retreating sensations and a repulsion that makes you sick. The sensation leading up to the vomiting itself is intolerable, and it occurs because of a past, present, or even future event that the individual is focused on and feels sick as a result of it. The body doesn't know anything about food. The individual "self" throws up the food because, according to its personal taste, it is interpreted as causing repulsion and sickness, not because there is anything really wrong with the food.

After a person has some tasty food to delight himself with, he is told that the food was old and rotten (though it wasn't the case), or that it is forbidden to eat. He would surely throw it all up in response if he only feels nausea and extreme rejection to that food. This sensation is an untamable sensation that can be ordered by will power. When we eat "actual" rotten food, without a sense of nausea, we can't throw it up by will (the body and mind have no wills, diagnosis or sense of judgment). The result in this case will be a poisoned body.

Side effects, like gastroenteritis, will occur by the "self" according to its toxic sensations.

On the other hand, there wouldn't be a need-based, immediate purging. In conclusion, **throwing up is a result of a sense of sickness and utter aversion to a certain condition. This is an individual situation, which can also occur when our taste buds reject a particular type of food. This is another sensation of regression deriving from the "self," according to its personal interpretation.**

DARWIN'S TRAGEDY

Richard Darwin returned from his excruciating journeys in hopes of writing and spreading the information that he carried. But at that moment of truth he fell ill with a mysterious illness, which he described in his writings. The ultimate question is: how come he survived his travels and journeys in places infested by many diseases (like malaria and others), and came to his demise in his peaceful home? Explanation: we return, yet again, to the urge. Darwin embarked on his travels due to an unleashed urge, moving him to make discoveries. In this relentless condition he felt strong, and thus transcended himself, and kept himself up above the different ailments. But when he had to return to write, he was faced with a block that he wasn't expecting when his discoveries were dismissed.

Let's enter his "self" for a moment and interpret it with words: he had an urge to spread his discoveries and the solutions he discovered after agonizing efforts. But there was a block in front of him. That is, urge versus block. It's a sense of a frontal collision which, in itself, is in an intolerable condition of his "self." The situation was a "short curcuit" of sensations that resulted in a catastrophic lack of conduction. That's exactly how his body behaved. According to the description in his letter, he vomited insufferable acid (that is the acid that the "self" produces by its consuming sensations). The view was blurry, as all images seemed to be in a black fog. He did not enjoy them. He became senile. His memory was lacking, he suffered chronic weakness, and chronic nausea.

I have chosen this extreme example to demonstrate the design of Darwin's infinite "self" that is a sort of "short circuit" created due to the burst of his sensations. Of course, in accordance with his condition, the typical resistance collapses and dormant ailments like malaria, typhus, and meningitis tend to break out at that moment.

Our lives are carried out by urges or the lack thereof. For instance, a football player plays in extreme weather conditions. Rain, cold, snow, cold winds and heat. Under such conditions, a man without an urge to play won't be able to survive and resist the various illnesses, unless he feels "elevated." You can claim that the player is running and warming up. If that's the case, what about the fans? They sit in the balcony without any kind of physical activity in stormy weather. The individual "self," therefore, along with the urge (the devoted fan), merge the individual with their beloved team, and manifest themselves against the severe natural disturbances.

During a game, the fan's sensation is mixed with an urge to win, in spite of the fear of losing. This mixture corresponds with the frantic interior sensations felt at that moment, and so the body reacts with excitement and fibrillations.

Once the team wins, the fear of loss is replaced by a feeling of winning, sensed by the "self." This condition resembles a spiritual elevation together with the feeling of free conduction. This sensation, though different from the previous one, still causes the body to react. May I remind you that the fan is not attached to his team's players. And all the drama has nothing to do with survival at all.

Another example: a man goes into prison for a week. His whole world comes crashing down because he doesn't see how he fits into this situation. And that is the equivalent of a humiliated, forced "self." Unlike him, a different man goes into prison to serve a life sentence. He is all chained and yet has a cigarette in his mouth. *This* prisoner is filled with pride and physical immunity, and thus behaves like a lion. If we were to interpret his body language, it would say: "I am a dangerous man and everybody should watch out for me. I am so dangerous I'm being chained. I'm "up" while you are "down." That is, of course, consistent with his personal interpretation of his "self", who was trapped in a situation of "no choice". The body of that prisoner will act in accordance with his sensations.

"MIGRAINE" — HEART AND HEAD ACHES

Headaches occur when the "self" suffers "headaches" due to individual and particular situations. When a man is experiencing stress and hardships, he carries them up to a certain limit without any given solution. At this limit of "no solution," the "self" will feel a sensation of "headaches" and physical pains. When the pains begin to frustrate, the man suffers a chain reaction ending in collapse. That is not a blow to the head. That is a combination of insufferable situations that manifest themselves as headaches by his personal interpretation. In this extreme state of pressure, stress, and helplessness, the heart loses its flexibility according to the traumatic condition. As a result, pressure is created within the heart and a lack of conduction occurs within the blood stream. In some cases a hole in the heart is made to release the pressure.

And on the same subject, what about genetic heart defects? Why are certain children born with a hole in their heart?

And again: man wasn't created as he should have been. Man was made by what happened during the evolution. There were always hardships. And in this case, the heart was struggling throughout evolution according to the sensations of our ancestors. Everything taking place during evolution, whether for good or bad, is sealed in our genetics without the guidance of a helping hand.

Unlike mankind, there are no major defects among animals in nature. That is simply because nature has only one option – being effective. Animals with defects do not survive, so the defect isn't passed on genetically. Let us add that nature has the freedom of expression, unlike what happens with the defeated mankind, especially among children and women, who are forced to become this way by the dominant male.

The various heart malfunctions aren't the same for everyone, since each person has his own individual sensations. For instance, a particular "self" will "eat" its heart out because of a certain problem such as a person who might have been deceived or robbed of all his money and savings. That "self" doesn't "eat" its heart out and will actually feel distress in the heart. And those aren't just plain expressions. The heart will actually operate according to this person's lingering sensation. Unlike him, his father, who is sorry for him, will also feel distress in his heart, but a different sort. His would feel like an internal stinging in the corner of the heart or a cramp, a distortion or any other sensation. The stinging in the heart could, for instance, be "consuming" one of the valves.

Let's carry on with the father and son. Say the son had all his money restored to him and his robbers even apologized to him, but he neglected to tell his father. The son's heart is, therefore, relieved of its earlier sensation and from now on will carry out depending on his new sensations. But his father's heart will keep on acting according to the previous situation. Again, the brain can't and won't interfere with the sensations of his owners. Those are strictly individual sensations. There is no intentional interference of a mediating brain. If the brain had any understanding, or if there had been an entity that would have designated brains for such a purpose, then the body – in this case, the heart – would operate and act perfectly, apart from the sensations of the "self." Meaning, the brain would have activated and operated the heart as it should have or as it was made to do by its maker, with no relation to the owners' sensations. When the "self" feels headaches, these are physiological headaches that occur in spite of the brain's lack of nerves. That is the "self" determining things for better or for worse.

A footnote: Once you, the readers, infuse into this book such aspects of previous knowledge as: "It is a well known fact that one should avoid anger because it has a negative effect on the body," this book becomes misunderstood! You should read the book from cover to cover, and only then combine all the missing pieces together and add them up. This is the only way to gain a pure understanding of the principle of our creation.

As for anger, it is not made in order to achieve anything. Anger is not individual. The "self" is upset when it finds situations to be illegitimate and frustrating, by his own interpretation. That is a heavily weighed baggage, ready to set off to achieve some relief.

Blood pressure

Blood pressure is high when an individual is undergoing sensations of stormy emotions, anger and unfed frustrations, stress, etc. Blood pressure is low when an individual loses hope, feels desperate, humiliated, or detached from holding on to life, and is exhausted and weak. This is the sensation of "I am going downward."

It should be noted that lab tests are not necessarily an indication of an individual's health condition.

On one hand, there are cases where lab tests show balanced

results, when the individual really feels balanced and his sensations are reasonable and positive.

On the other hand, there are also cases when lab tests show balanced blood pressure, when actually, that individual does **not** feel balanced at all. He feels and senses a mixture of all the above.

CHRONIC FATIGUE AND WEAKNESS

Chronic fatigue and weakness are the result of mental exhaustion originating in desires that are due to the fact that various blocks remain reserved and are unexpressed. It may also stem from long-term humiliation and a sense of being stuck in place. Alternatively, it could also be a case of carrying out the phenomenon, such as with chronically worried parents who pass their morbid sensations to their sympathizing children (see **"Carrying out Diseases"** episode). Yawns and sighs are an involuntary reflex of chronic fatigue and weakness. In extreme cases, we run across serial yawns, accompanied by a sensation of extreme fatigue, and yawns in the state of frustration. That is a dark sensation.

Feeling worn-out is an individual sensation where the

"self" is "down" for quite a while. It is aware of the situation and as far as it's concerned, there's no light at the end of the tunnel. Therefore, the "self" is feeling as though it is slipping away. It slips back and forth from the top of a slippery, tall, and steep mountain and has nothing to grab on to.

The exhausted "self," that is weak and lacking energy, equals an exhausted body (and brain), which is also weak and lacks energy. The deeper you sink, the grayer your cognitive thinking gets.

In this deteriorated situation, certain ailments such as fungus find a comfortable living space within the surrogate body (i.e., the inferior and passive "self"). This phenomenon occurs due to the "Scales rule." Neither fungus nor the bacteria are in charge of chronic fatigue and weakness, as some may think. Keep in mind that the microscopes lack awareness and insight, and as frustrating as the diagnosis may be, they do depend on individual interpretation. Germs, for instance, are not racist and do not choose their victims out of personal considerations.

Recommendations

You should gather your strength and start living again by changing the atmosphere, the residence, and the company.

You should nourish the downtrodden "self" and empower it to once again become the proud "self." You should change your body language to be more optimistic. You should take up a responsible and creative field of interest so as to set your mind straight and get rid of the past. You should carry out your hidden and blocked desires (like getting rid of a demeaning partner). **You should carry out the change with determination as you "shovel" your way out.**

Chronic illnesses, such as inflammations, general weakening of the body mass, aching joints, sickly looking skin, poor digestive systems, or bad memories, all of which were plaguing the prior "self," will disappear as though they had never existed as the new and vibrant "self" appears.

DEPRESSION

Reminder: all laws of nature coexist in the animal kingdom. Whether it is man or the rest of the animals, the laws of nature are not to be taken for granted. Our brains don't manage our lives on their own. **Neither we nor the animals operate by the amount of substances that are in our brain and body. Not at all.** All living creatures are a hundred percent living souls, alive and walking, without any external interference to our need of running our lives and our bodies. This fact is like life and surviving in the "quicksand swamp." According to this insight, the sense of anxiety is completely legitimate; and survival, therefore, is not just a formula for how to get food, water, or having the ability to fly. It is much more than that because the survivor has a sense of grabbing onto life, which makes him happy. That is the struggling "self." longing for life. Without this condition there is no basis for the making of a biological body.

Depression among animals or humans **isn't a result** of "short circuit" in the brain's serotonin level. The process is reversed, as a matter of fact: the serotonin level in the brain drops **in accordance with the sense of depression.** The depression is caused when a man sinks into difficult situations. These situations are subjectively difficult, depending on the individual's sensation. It could be unrequited love, loneliness, desires that disappear into the distance, a feeling of loss or of not fitting in. Sometimes it is a mixture of these situations. Detachment from society (by a real or initiated social boycott) is one of the causes of this sense of "sinking." According to the "Scales rules" that I previously mentioned, a man's awareness of his situation, where "happiness belongs to others," only causes him to sink deeper into his depression. Since the body "behaves" and conducts itself according to the "self," the person's serotonin level drops and rises based on the condition of the "self."

It's not the serotonin's role to take the individual out of his current situation. The serotonin won't change his loneliness or desires that slowly disappear into the distance. If a battered woman is given serotonin, yet her violent husband keeps abusing her, either verbally or physically, she will still feel humiliated, despite the administering of the neurotransmitter. Moreover, **the serotonin level in her body will fit the sensation she experiences at that current time.** That is a legitimate level of serotonin to have in your body. Therefore, taking the substance could cause side effects, since it creates a kind of "short circuit" (or collision) between an artificially, non-legitimately fed body and the woman's humiliated, and in her case, legitimate sensation.

To demonstrate this principle, imagine a child crying because he is in some sort of distress. His crying is legitimate in his mind. But if his brother makes fun of him or his father shushes him, he will feel a type of "short circuit" of sensations, since he can't express his crying instinct. This is a frustrating situation. That is the principle of serotonin.

Adding a low dosage of serotonin might provide some support when the cause of the depression isn't tangible (such as the case of sinking into apathy).

As for serotonin itself, the question we should ask is: is serotonin the only positive substance in our bodies?

This book aims to generate change in the psychology field. Psychologists have a lot on their plates – changing a certain "self," playing with it and remodeling it, charging it mentally with energy and "pulling it out," if only a bit, from the "quicksand swamp." All these might help change their interpretation (the treatment should be "rolled and pushed" forward so as to prevent the patient from becoming bored, and thus further depressed, and also to make sure that the patient doesn't become too fond of his own disadvantaged state and become addicted to it). All of this is because we are born with brains and awareness, but with no instructions. Each living individual is actually alone, without a specific entity running his life and taking care of him. That's why living creatures are vulnerable in the face of depression.

Depression is accompanied by a sense of loneliness, **despite having friends around.** This is a sensation that is manifested by such thoughts and phrases as: "Life is for others while I sink and drift away."

These sensations manifest themselves specifically during joyous occasions, which the individual considers as "belonging to others" or during a "sad" mood, which the individual interprets as him coming down with something. Such sensations will eventually lead to an actual cold.

Depression is therefore a sensation of the "self" that matches a certain individual situation. You feel disconnected from life, in one way or another, and helpless.

Adding serotonin, when the diagnosis is not exact, can thus, only cause dangerous side effects. There is no black or white situation.

Speaking of black or white, let's take a girl who is hurt and depressed because the other children in school call her "black" and "the N word." If a person with authority explains to her that black is beautiful and she is indeed convinced, her hurt feelings will improve. In that case, the child might still be hurt, but she wouldn't necessarily suffer from depression. This is yet another example of changing the sensation by altering our interpretation of the situation.

If that girl had self-esteem, she wouldn't have gotten hurt, since her self-esteem wouldn't depend on social perceptions.

When self-confidence is nonexistent, the individual's body language invites others to feel "superior" to them and treat them as a punching bag. But the moment self-confidence appears, the body language signals, "I am not the person to release your frustrations upon."

Recommendation
Our lives should be filled with activity, to engage the lonely and insecure "self." We should engage in social and lively occasions, in order to feel in the center of the crowd. Each of us should join groups and activities that encourage the "self" and elevate it by engaging in activities such as dancing or even rappelling; thus, inducing laughter, personal empowerment, change, self-esteem, and self-fulfillment. This is done to empower the sense of one's ability and self-confidence. We should all "push and shove our way" into the center of a social atmosphere against all odds. Be aware that every living creature instinctively longs to "climb over" others by taunting, insulting, and aiming arrows. Understanding this principle will release us from the sense of burden and might push us upwards. Remember that a sense of belonging is the basis of life.

HAPPY PEOPLE — DON'T THEY GET SICK?

A person who is happy has a better operating body with a much more effective metabolic activity, a faster metabolism, and a "freer" immune system; along with a steady flow of vital substances in the bloodstream, good connection of the joints, and a better conduction both in the blood and the digestive system.

But some question remains unanswered, one being: happy people – don't they get sick?

In order to explain this, let's take a walk in the jungle and try to imagine ourselves as the "self" of the animals. And, may I repeat, we are in *the wild*. Animals in the wild are standing alone in the face of numerous dangers and struggle for survival from the very first moment they are born. Having no other choice, the animal is aware of its body. It can't choose this or the other, nor declare anything. The altering situations are the ones dictating to the animals how to adapt to their bodies. They live beneath the open skies in extreme conditions, and so they are constantly alert to the threats around. The animal sleeps with an eye half open, aware of the efficiency of the body that is ever ready for a possible retreat. When it is in the savanna eating, it doesn't relax the body in hopes that everything will be all right, but is always alert and ready to flee from predators. In this daily situation the body becomes more and more invigorated, since the animal never knows where the predator will attack from. All its bodily organs are on instinctive alertness: will the threat arrive from the right, left, back or front? This prepares the animal for a quick retreat and even for a "do or die" battle. The animal is always instinctively aware of the physical surprises it might encounter, like burrows and stones concealed by bushes, fallen trees, or any other obstacle that might stand in its way and determine its fate. The animal can either intuitively resist those obstacles, or be tricked by them and pass as "prey", literally, by slipping or breaking a leg. In that situation, the animal's "self" is aware of its efficient body and becomes one with it.

For humans, however, these laws of nature have been altered. Man has been transformed from the ultimate survivor, living in impossible conditions and with effective senses, **into a feeble creature, disconnected from these natural conditions.** As a result, man is less in tune with and less aware of his body. Man carries his body with him, letting go of its efficiency, and so the physical immune system suffers and the individual becomes more prone to disease. At times, this might lead to a complete lack of resistance and a general state of conformity, which is what happens when the senses aren't used adequately and are lost. In this situation the body becomes completely distorted, because the human body doesn't have another owner but man himself. In that case, the designing "self" is disengaged from its body.

Beliefs like: "We are a product of a creator and the fate of our bodies lies solely at his mercy"; or rather "We are a genetic, biological machine, and our bodies are controlled and operated by our brains" are inherently wrong!

When the person's line of thought is based upon the belief that his body operates and is controlled by external forces that are beyond his own control (like genetics, hereditary, a virus, luck, fate or a transcendent maker), he lets go of his body. He is no longer attached to it; and as a result, his body is distorted.

Snoring, for instance, occurs when a man is sleeping and is unafraid and, therefore, not alert to any possible danger, but is rather completely at ease throughout his sleep. The relaxed, sleeping body is disengaged from the formative "self," which is what causes the distortion. The same principle applies to body fat, when the body doesn't have a formatter, for reasons that were already mentioned, like a potato bag. Diabetes is one of the illnesses that may arise from this phenomenon.

BONE MASS

Science and biology experts endlessly claim that bone mass starts to dissolve from a certain age and that we should therefore fight osteoporosis. One way is to consume calcium-enriched food or by adding vitamin D to our meals.

Absolutely not! These conclusions are drawn by old, dried up surveys that cater to a decreased range of ages.

Bone mass formulates through the unstable and infinite sensations of the infinite "self." I have mentioned deep sleep earlier. The bone mass will conduct itself according to the sensations of its owner (the "self"). It is more or less condensed, depending on the person's level of alertness. But that's not all. Each person has a different "self" that changes, according to its relevant situations, the way one senses and feels. Does this person feel strong? Is he militaristic in the face of strenuous difficulties? Or rather gentle in character and unfamiliar with everyday hardships? And maybe he is self-aware, for better or worse. Is it: "I have very strong bones. I feel strong" or rather "I have weak and crumbling bones"?" Of course, these are infinite, individual sensations of the infinite "self." be it temporary weakness or chronic weakness that is derived from boredom, or forced boredom and wasted time (due to lack of active planning and so on). All these possibilities are a combination of individual sensations, which cause frustration.

During sleep, the individual feels the absolute passivity and limpness of his mass. Therefore, the bone mass structure will be the same and react in sync with the sensation. Let's take the tiger that sleeps lightly and alertly, unlike the human being. It is attentive to every rattle, even the slightest of which would quickly get it prepared. Therefore, its bone mass is highly condensed, relative to the bone mass of a human asleep.

If there were an entity in control of the human body, then the bone mass would have been continuously dense. **No creator would have an interest in creating a body in which the bone mass drops during its sleep, during a state of rest, or lack of physical activity.**

The bone mass of the aging man corresponds to the

state where the old "self" feels worn-out, exhausted, and diminished like his variable individual sensations.

The food we eat contains a lot of calcium. Its surpluses are irrelevant to our bodies and are passed on in accordance with the individual sensations of our "self." Adding calcium only burdens the cleansing systems of our body.

Osteoporosis takes place according to the "quicksand swamp" principle. Meaning, a woman in a certain situation is aware of her own fragility. She embraces this situation passively and with retreating steps until she eventually collapses. No entity or remedy stops this occurrence other than her "*I am my own maker*" (See "**Resistance**" episode).

I hereby declare: there is no food, potion, or remedy that would strengthen our bones. There is only one alternative: to imitate nature, to feel like a strong, surviving warrior, in addition to mental treatment.

In conclusion, bone mass is not stable. It can dissolve in a flicker of light in a state of fear and sudden shock, or for a long time, when we recede from our grasp on life.

Bones and the Sun

Is it the responsibility of the sun to strengthen the bones?

If you think the answer is yes, then let me ask you this: what about the mole? And the polar bear? The man who lives at the North Pole?

It can be said that the bone mass is determined by the infinite, unstable "self."

The wrong information that men possess, which says that the sun builds up our bones, has led to twisted behavior. For example, there are those who leave domesticated animals outside, in the heavy desert sun, because of this belief.

If extra calcium were to solve the problem of osteoporosis and bone density, then the issue of bone mass would have been solved and I would not have written this book.

Does a lack of sun rays cause depression?

Not directly. Indirectly, an individual might feel down during a long period of a gruesome, dark winter because he longs for the sunlight and the vibrant atmosphere. When a person is down, their bone mass dissolves accordingly, in correspondence with their deteriorated state and the "dissolved and lacking of mass" sensations that they are experiencing.

That's why a decent portion of sunbeams might contribute to the bone mass, but in an **indirect way.**

SWEAT

Can sweat really cool you off? Why, sweat is hot and sticky. What about during times of stress? And did the brain, or the creator, or evolution ever think (in advance) that perhaps the sweaty individual should run or live by a river, or any other source of drink water, in order to regain the fluids that were lost to the body? What about the salts that we lose? What happens during droughts or during the summer in the desert? The answer is: **sweat is an instinct of the "self," for good or bad.** The "self" is hot, the "self" produces water. The body obviously responds by the warm sensations of the "self." The more the individual feels stressed about any form of heat (not necessarily hot weather), the more his body produces sweat **even if it's dehydrated.** A man can dehydrate even though water is present. It is a sensation of dryness and exhaustion ("the quicksand swamp"). These reactions occur without any logical conduct. Sometimes an individual will sweat when it's cold. That is, when he feels a certain stress, which he interprets as heat. There are also cases when a living creature shivers as a result of feeling cold, when the weather is not especially cold. This happens when his sensation is kind of chilling sorrow and cool atmosphere. This wasted mechanism is not used to get warm when one feels cold, or cool off when one feels hot. This means, again: there is no entity that has created well planned and well programmed living creatures.

Conclusions

The sweat glands were created by the sensations of our ancestors, as a reaction to heat, without any consideration of the above-mentioned conditions. Everything that took place in the evolutionary process is sealed in the genetics for future generations, whether we like it or not.. To summarize, in the absence of cool wind, hot, sticky sweat doesn't cool you off.

Take, for instance, the snake, which learned how to preserve its body fluids due to the "no choice" process.

PREGNANCY AND HORMONES

Pregnancy Problems of Conception

A primal example: a mare that mated with a castrated horse, believing that she is pregnant keeps a protective distance from other males. Her stomach becomes swollen and by the end of the term of pregnancy her udders produce milk. She is then examined by a veterinarian, who, to everyone's surprise, announces that she isn't pregnant at all. This phenomenon is called "false pregnancy" and occurs with humans as well.

A different example is a newlywed couple who got married against their will. They tried to conceive, but in vain. The woman went through different fertility treatments, to no avail. Knowing her fertility is damaged, her body reacts accordingly. Years pass in an obsessive race toward the desired destination. By the eighth year, when she has already given up on the excruciating race, she discovers that she is pregnant within three months.

Conception too is a "no choice" process. The "self " longs for energy, life, continuity and dividing. Since we are not reproductive machines, **the sensation is the condition for conception (continuity and dividing).** A woman will become pregnant only if she feels like a mother-to-be one hundred percent, without any disturbances to thwart this sensation, like stress or various fears.

Some people believe in all sorts of "magicians." When such "magicians" happen to succeed, they are thought of as powerful magicians, but when in all other cases they do not succeed, then they are forgotten. If a woman was to turn to a miracle worker only as one last attempt (after visiting several miracle workers), chances are it won't work, since she understands there's not much hope left. But if a woman believes with all her heart that this "miracle worker" is the ultimate salvation, then her "self" transforms from a "self" that experiences fertility problems to one that is a hundred percent fertile. And so it shall be according to her personal interpretation (see "**Placebo**" episode).

On the contrary, a woman can conceive, even if she doesn't want to, because the body doesn't operate by will but by sensations. So if she convincingly feels that she will become pregnant, then she will, despite her feelings about it, according to her feminine sensation.

We should thus let go of our fears and just flow with the current. A woman should avoid giving herself an ultimatum. The pregnancy will occur naturally, without anyone pushing and setting deadlines. Different types of stress distort the reproductive system. The worrisome awareness about the defective reproduction, so to speak, leads to a sealing of the distorted reproduction system in the future. Thus wars emerge against windmills. This is the "quicksand swamp."

Everything the "individual self" is expecting, or convinced of, comes true by self-fulfillment, for better or for worse, like not being able to get pregnant, impotence (not necessarily sexual), or any other failure.

Another example of this is milk. Milk isn't homogenous in its ingredients, its quality, and its amount of production. These ingredients are measured by the (fickle) mental state of the mother and how "motherly" she feels. The quality would only deteriorate in extreme cases, such as if she feels repulsed by her infant, or if she is experiencing mixed feelings or mood swings.

Hormones aren't designed to determine the structure and qualities of the milk. The creator of milk is the mother's "self," affected by her internal, infinite, unconscious and unstable sensations. As a result, the digestive system of her infant is affected by the mother through its own interpretation (his individual taste).

Pregnancy anxiety is not an illness, even if it *is* the sensation experienced throughout the pregnancy or post birth. Seeing anxiety as a legitimate condition, as far as the "self" is concerned, we should examine what is happening within the woman's "self," in the depths of her soul, before passing judgment. For instance, she could be experiencing such fears as: the fear of letting go of youth; the anxiety about the birth being traumatic; fear of bringing a child into the world at a young age; fear of not enjoying life to its fullest. Or rather she could be worried about what will become of the baby. Can she provide for him? What about when he grows up and moves out of the house, leaving her all alone? Or, did she ever want this relationship? Does she feel a part of this situation? There could be an entire array of contemplations regarding both good and bad sensations, mixed with frustration.

The same principle applies to animals. The fetus of the mare, for example, is either rejected or absorbed, without her awareness, in times of stress and mental anxiety. This is a sensation that contradicts the maternal instincts (continuity and dividing). That is the infinite "*I am my own maker.*"

In conclusion, we are not machines operated by hormones. We should immediately let go of our hormone theories and accept that man is the slave of the information he has acquired and of the fixations he adapted to himself.

Testosterone

The testosterone level of the male (be it human or animal) is determined by his sense of either his own "alpha male" pride; or by his humiliated, shattered, and broken sensation of the "self." Testosterone doesn't have any direct relation to sex. Take, for example, a horse mating with an entire harem of mares. His sense of pride transcends beyond belief and his testosterone level skyrockets. But if it is suddenly disturbed by another horse entering into its territory, dethroning and banishing that horse, then its testosterone levels would drop according to the sensation of degradation the horse experiences, and it would become exposed to various ailments. As for his successor, his testosterone levels would rise, with respect to his sense of triumph, dominance, and pride. On the other hand, the testosterone level of a castrated horse would correspond with his sense of "self." either of a proud horse, or a humiliated and damaged horse.

The infinite interpretations match the determining "self," according to endless interpretations and infinite situations. If a child is slapped on the face in front of his friends, his body will crash based on his humiliated sensation. A boxer, on the other hand, would choose the battle situation, being fully aware of the beatings he would get. His testosterone levels rise along with his victory in relation to his skyrocketing sensation. But once he hears words that he **considers** insulting, even this muscle-packed boxer would fall apart, causing his testosterone levels to drop according to the new interpretation of his "self."

In one case, a horse has lost his desire to train and race (because of "human" ignorance). Since it was treated like an emotionless racing machine, it was constantly injected with testosterone to induce its passion, over and over, in a forceful manner. As expected, the horse's condition worsened. It was treated by different vets and caretakers until it was finally labeled as "not a good horse" and it was forsaken. The testosterone administered to it is like adding serotonin supplement to a humiliated and beaten woman… The science "sprints ahead," but in the wrong direction.

Forcing a proud horse by pulling its ear in the trailer before a race is utter humiliation and a recipe made by stupidity. The testosterone levels will drop and dissolve, according to this humiliating state. The lost man is convinced that animals are biological machines that exist solely for his purposes.

Again, the same principle applies to humans. For example, the testosterone levels of a proud child who is insulted by a random comment uttered in the presence of his friends will drop and even be absorbed according to his humiliated sensation and individual interpretation of the event. All this occurs despite the fact that this drama had no physical activity or biological interference. It is the determining "self."

I should mention that a male (human or animal) is sensitive and emotional about his sensations of pride…as well as his humiliation. For instance, a human male finds it hard to get advice from another man, even if it is reasonable and constructive advice. This is because, according to the "Scales Rule," his sense of pride is shaken. I refer to statements such as, "I know better than you" or "Don't tell me what to do." This type of behavior causes a distorted and wasteful fixation, and all that implies.

B12

One of the roles of this particular vitamin is to provide food and energy to the red blood cells that produce oxygen to the body. Science and biology have, therefore, concluded that adding this vitamin in vitro nourishes the red blood cells and enhances the biological performances. The internal sacred order calls again to all you scientists and biologists to dispense with your dependence on labs. Despite the fact that this conclusion was never proven, this blind belief carries on to this very day, because "a belief is a belief and one must never question it." (Further explanations will be provided throughout this book).

The same principle applies to different vitamins like iron and selenium. Anemia, for instance, is a kind of deteriorated, "anemic" sensation, during which the individual "self" is in "the quicksand swamp." Adding iron is missing the principle.

These beliefs led humanity to the certain assumption that animals or humans are biological machinery. This is the source of our forsaking logic and for the agony of animals – and humans as well.

ADRENALINE

Adrenaline is a substance that is created by the body when the individual thinks or feels a sense of "action" or an emotional rollercoaster, induced by a past memory or a situation in the present, or even an occurrence in the near or distant future. This occurs according to their focus; their thoughts; and their interpretation. From that person's point of view, the "action" sensation is vivid. It is a sensation that equals adrenaline. The substances that are created as a result of the "action" sensation in the body, for example: an attacking predator wouldn't be perceived the same way by all the living creatures that are being attacked.

An animal that is backed into a corner and is paralyzed with fear will secrete different substances than an animal that is running for its life. Yet it maintains its self-confidence and pride.

The potential prey cannot ask its predator to hold on until the substance starts to flow through the bloodstream, reaching our entire body. Therefore, the substances in our bodies aren't made for or against our benefit. Evolution didn't take care of that. Substances are simply created because of infinite sensations, not according to any logic or order, but as a mixture of substances spread through the body according to its owner's infinite sensations – for better or for worse.

Moreover, **the brain doesn't prepare the body for an independent action. It is the infinite "self" that is either frightened or not, depending on its individual situation.** It is either prepared or unprepared, reacting or fleeing for its life, on its way to a dramatic adventure that according to its interpretation is about to be either successful or a failure.

The brain doesn't produce adrenaline to numb the pain (that is a basic misconception). Had this been the case, there would never have been patients suffering from such ailments as chronic pain. In some situations, the individual is so focused on a particular dramatic event that he becomes completely detached from any physical pain. And there are other situations in which the pain is dimmed, when the dramatic event is positive and nourishes the passion. On the other hand: pains burn when the drama is traumatic (by individual interpretation). Endless hormonal changes are constantly occurring in the body of each living creature. These occurrences are run by the owners of the individual (the "self"). For example, while watching a thrilling movie the viewer's body experiences plenty of physiological changes, according to their sensations and by their interpretation. The brain doesn't differentiate between imagination and reality, and holds no interests of its own.

In conclusion: it is the "self" and its infinite and individual sensations, according to its interpretations. The brain doesn't recognize predators. The brain doesn't prepare the body for an attack or retreat (much like the widening of the nostrils). The brain doesn't perform any act in an independent way. In the worst moments of fear, all bodily functions collapse according to the individual's interpretation. This goes against logic, will, and need.

As for the nostrils: they twitch or widen according to the sensation – which is the "quicksand swamp" effect.

TEETH AND EYES

Teeth are a dominant organ among animals in the wild. The animal is aware of them, feels their strength, and even shows them off (in accordance with the "no choice"). Unlike them, humans, who are able to manage without any teeth, and possess an unconscious choice in the matter, neglect the quality of their teeth and abandon them passively. Again, if there were some sort of guiding or creating entity, then teeth would have been strong and self-renewing (more than just). But what can we do about it, now that it has been carried out throughout evolution and sealed into our genetics? This phenomena of human deterioration is not at all related to a lack of calcium, fluoride, or any other theory.

Seeing as in this case **we should replace the currently**

existing "self" with a "self" that has teeth as strong as a vulture's teeth, imagine them tangibly, and they bite through bones. This is especially true during times of crises or hardship (which project to the teeth). Don't let that snowball roll down that slope.

The eyes are dominant organs in nature and without them death is only a matter of time. In order to improve our sight, we must sense the change in our interpretation: "I'm a hungry hawk; my chicks are hungry and waiting for food. I soar above and capture every movement made by a tiny rodent down by the bushes." And of course, you must sense the situation sincerely and turn the "self" with the bad eyesight to a "self" that has the eye of a hawk. We must always rise above the existing situation, and see clearly and relentlessly. It is possible when the sensation is not dark.

An extreme example of poor eyesight that is withheld from our consciousness is welding. When one welds metal using electrodes, a focal fire is created. My question is: why do a welder's eyes burn up to the point of blindness once he looks directly at the fire without a mask, even from a distance? After all, the fire is sufficiently far and doesn't have the effect of heat (this point can be proven by a laboratory experiment on a dead animal. If the animal is dead, its eyes no longer have "an owner," and so a simple exposure to fire will prove that it won't affect the animal's eyes). If so, it is a focal fire that scorches the owner's eyes. The eyes are not directly affected; rather it is the "self" burning his eyes with the searing look. The eyes don't strain themselves and get tired after an extended reading. The eyes don't see and they don't "spice up" reality on their own. It is the "*I am my own maker*" that sees through them, for better or worse.

This paragraph is dedicated to those skeptics amongst you. Some people undergo eye surgery to replace the lens. The eye is operated upon and improves in quality as a result. At the same time, the other "inferior" eye also improves and becomes equal in its quality to the operated eye: all without undergoing any surgery. This fact proves that the quality of the eyes is determined by their owners. That is, it's the "self" that sees well after the surgery. A new sensation carries with it an improvement in the twin eye's quality.

BACK PAINS

The body works more efficiently when we do the things that we have the urge and passion to do. **When a man lifts**

an object without feeling the urge to do it (desire), his body reacts appropriately to his sense of "self." In this case, the back isn't prepared for the lifting and for its owner's resistance, and as a result it suffers injury and pain.

The brain, a plain, biological computer, doesn't prepare the back to complete this task.

Chronic back pains are also caused by mental distress. Carrying that weight in daily life is a sort of mental burden that dissolves the back's mass. This is a result of feeling as if you are slipping away from your destination. The "self" repeatedly feels that the road to its destination is slippery, and far from grasping or fulfilling. This would normally characterize an individual who feels as though he is carrying the world's burdens on his shoulders. This is felt when, for instance, one is desperately looking for work, or when doing one's work and resenting it, or even when one is experiencing an ongoing bankruptcy.

And sometimes the individual "self "senses that its body is only imposing upon it and thus becomes detached from it (see "**Resentment**" episode). These are the main reasons for back pains. In these cases, we must never trust the abdominal belt because if we rely on it too much for support, then our backbone will weaken (both mentally and physically).

Recommendations

The individual must encourage himself and make plans to get out of the given situation. At the same time, the sense of resentment must be activated using instinctive and tangible imagination, by contracting the back and stomach mass, and not backing down in a passive and surrendering withdrawal, accompanied by the release of mass and the relaxation of muscles. Back pains are usually followed by other hardships in the digestive system and a load is created on the lower organs.

Scoliosis usually appears in individuals who literally have a "spineless" nature and passive attitudes toward their own bodies: almost like a snail supported by its shell. There is no potion that will straighten one's back. The only thing to do is leave the safety and comfort of our bubble, and live independently in the rugged atmosphere of the real world. That is done to activate the lost resisting sensations (see **"Resistance"** episode). There are various ways of encouraging oneself and one's body to do so, and restoring it to its designated owners ("*I am my own maker*").

THE CONCEPT OF TIME

The ever slippery concept of time is a hidden law of nature, but I have solved its mysteries. Comprehending might be difficult, but before we discuss it I would like to play a questions game with you. Take blank pages, envelopes, and some writing implements, and hand them out to a group of people. I would like to ask you to answer each question and place it in a different envelope.

My first question is a theoretical one: say the body of a ten-year-old dinosaur that has been frozen in Antarctica about 70 million years ago is found entirely preserved. Let's say that with the aid of modern technology we can defrost it. Suppose we do defrost the body and the little dinosaur returns to eating grass; how old do you suppose this creature would be today? Please close the envelope after giving your answer.

Now let's assume that 70 million years ago, a day had only twenty-one hours. At that time, a female dinosaur laid two eggs (that were simultaneously fertilized with a living fetus inside). One of the eggs had hatched, and the little dinosaur lived for ten years and then froze to death. The other egg, however, didn't hatch, but remained frozen to this very day.

And now, let's assume that they were both defrosted today

(the ten-year-old frozen dinosaur returning to life, and the frozen egg from the hatching day). Both are coming to life today. As a result, the dinosaur from the defrosted egg that would hypothetically hatch today would look younger than the ten-year-old dinosaur would.

This brings me to my second question: how old is each one? May I remind you that when they were born a day had only twenty-one hours.

Let's take a simpler and more realistic example: a certain type of salamander lives in the North Pole. The salamander freezes each winter and then defrosts, coming back to life during the summer. That brings us to the question – if the North Pole's salamander had been born ten years ago (according to the time determined by humans), how old would it be now? Ten years old? We should add that the salamander didn't experience winters. It had never lived through a winter and is unaware of them. If, theoretically, the varying climate suddenly changed to frost for another 70 million years, how old would that salamander be once it defrosts back to life?

Please open your envelopes now, as you will probably discover the answers in the group are inconsistent. Each would most likely get confusing and inconclusive answers.

This is the first stage of undermining the so-called rule of "resolved time."

In addition, the dinosaur born 70 million years ago and coming back to life would probably feel slightly confused over the time differences influenced by his prior concept of a day being twenty-one hours, compared to our twenty-four-hour day and it would take the creature several days to adapt.

On the contrary, the dinosaur that has just hatched won't feel "jetlagged."

North pole winter time, sun is NOT shining. Two Eskimos run into one another. They want to plan a family reunion and future pairing. How will they do it (in an era before time was invented)?

Try a different exercise. This is an imaginary one: sit in front of two clocks (old-fashioned ones with hands, not the digital ones). One is set to the current time in your country and another to a faraway country's time, with a six hour difference. After ten minutes of watching, the watch with the current time stops and the other keeps on ticking. What is the time then? One individual "self" might calculate a six hour difference, while another would decide to run his life according to the foreign time, which indicated either the past or the future.

In fact, time is an unconscious and infinite sensation, existing according to our individual interpretations, and the endless situations interpreted by the individual "self". Meaning, the North Pole's salamander doesn't feel the sensations of time while it is frozen, because its body doesn't have any owners or a concept of time. The time during which it is frozen is only time as far as the living viewer is concerned.

The brain has a **so-called** biological clock, but it doesn't count our time, and doesn't tell us what time it is or what a day is, and it doesn't wake us up at a certain hour, based on our expectations. The "self" feels an individual time that isn't the same as the time sensed by a different individual. The biological clock operates by a time that the personal "self" senses, yet it is rather different in its principles compared to the mechanical watch that man has created. In addition, we must also consider the factor of fatigue that is blurring the sense of time (which is a blurring of the order of events).

We add up situations and events that we have experienced, and spread them over time intervals to remember or plan ahead; or we add them in order to learn and improve by not repeating the same mistake twice. All this happens in a subconscious manner. Mankind has used time as an ultimate and necessary tool, and even set it by data explicit to us: a watch. One hour. Sixty minutes in one hour; twenty-four hours; each time the Earth completes a full round. Let's stop here for a moment. If theoretically the Earth suddenly stops spinning, we would continue to rely on the watch, thus losing a day. In such a scenario, there'll be no tomorrow, yesterday, one month ago, last year, next year...We will continue going by the watch's time: one thousand hours ago, five thousand hours away (new measurements will probably be made in the future).

Let's say that by using a telescope, we could measure and even adopt the circular time pattern of a different planet. In this case, the watch's index will be problematic and we will have to reprogram it. Or maybe not: man dictated time by tools that didn't fluctuate. If he had measured time relative to the wind, then we would have been in trouble. We can't measure time by counting the spins of a windmill and running our lives that way, because the wind isn't stable and constantly changes.

For us time keeps rolling forward, even when our car drives in reverse, going back to where we came from.

The question of going back in time causes many philosophical discussions. A great grandson goes back in time and shakes his great grandfather's hand, supposedly before he was born.

If we understand that time is only a concept, then we can arrive at the conclusion that this question is irrelevant. Let's try a different experiment. Pour water from a rectangular bowl into a cylindrical bottle and freeze the water. Later, when we defrost the water and pour it back to the rectangular bowl, could we say that we have reversed time? No. This isn't relevant to the concept of time, but merely a process recreated. Not all processes in nature can be reverted.

Time isn't objective, but rather subjective and altering, according to our sensations.

Perceptions of Time

A man is on a plane, traveling at a speed of 300 kilometers per hour. The plane vibrates and makes noises. The view outside changes rapidly, while the "self" interprets time versus speed. But once a plane takes off, the sense of time by the "self" changes because there are no rapidly changing views, and there is no noise or vibration.

According to what was said, **the substance in the universe**

wasn't created at a certain point centuries or countless decades ago, but rather in the present and it exists without time. Time is the interpretation of each conscious individual who senses it. But if that's the case, then why do we sense time despite focusing on a silent object, or during boredom, and lack of activity?

Every conscious living organism thinks. The endless thoughts are running wild, subconsciously and uncontrollably. That is time. But let's say, theoretically, that we can freeze and stop the thoughts of a tested individual at a certain point. That individual will remain with one frozen image, without "self" scanning the thoughts. At this point, time will stop passing for his "self".

The substances of the universe go through processes and reactions, without which they would have been still forever and time would no longer be relevant.

In conclusion, **if the substance of the universe was still and didn't have a chemical structure, no reaction would have been made. Therefore, no planets, galaxies, and lives would have formed. As a result, there wouldn't be a concept of "time" and there would be no time measure** that would aid in determining how long ago the still, nonreactive universe was created. This means that the universe would be nothing more than a mist of still substances, without planets, galaxies...or time.

Another example: the material needed to create a wheel has been in existence for numerous decades (as far as we are concerned), but it took a few million years before man learned to create a wheel, and another few centuries before it was used for transportation. A wheel can remain motionless, but if it begins to roll, it indicates the beginning of the time period, according to the spectator's "self" and its interpretation of the speed of spinning. At this point, there are the materials that could be made into a wheel (as they have existed for countless decades) and there is the still wheel (that has existed for, say, a year). There is the spinning wheel, and there is the speed of the outer diameter or the inner diameter of the wheel. Again, these are all processes. Manufacturing a wheel is a process, but then so is the spinning of the wheel.

As far as the wheel is concerned, time is irrelevant. But there are other laws of nature that apply to it, like centrifugal force, G force, acceleration, friction with air, heat, and deterioration.

I will use another example to sum up the issue of time:

At **stage A**: we ask the tested subject to look at a still ball. His relation to the ball would be a subjective sensation of "time."

At **stage B**: we tell him that a fire is burning inside the ball. This changes his sensation of "time."

At **Stage C:** we tell him of another identical ball that has a circular engine in it. It's another stage that changes his sensation of "time." I should add that in this example, the sensation is unstable since the person doesn't know the speed of the circulating engine.

At **Stage D**: the person is told what the speed of the circulating engine is (say, one hundred rotations per minute). Thus, his perception of time undoubtedly becomes more coherent.

There are a lot of concepts that are just that – concepts. But for us conscious individuals, they are subconsciously tangible. Let's take a look at luck and randomness. They are not, in fact, materialistic, but rather an interpretation. The dice fall only according to the laws of nature. Luck or randomness only exists **as far as the conscious man** is concerned, along with the sensations of time.

In conclusion, time is nothing but an individual sensation experienced by the individual interpretation of the infinite "self."

The concept of time is not a substance or energy (like a magnetic force), nor is it empty. It is nothing, in fact. Time is nothing but a concept, and as such cannot influence substances. Only our sensations of time are the indication of processes occurring in nature.

As a matter of fact, there is no such thing as time. In accurate scientific language, time=nothing. Since we conscious owners say: "Time **is**" (or "magnitude **is**"), one might infer that there is such a thing as time. Hence, the language that we use is incompatible with the above scientific formula.

When was the universe created?

Past and **present** only exist for us conscious beings who have a sense of time. "Tomorrow" could theoretically become three more days from yesterday for us. But if a pilot flies around the world at a twenty-four-hour-a-day speed toward the sunset, that "tomorrow" will remain tomorrow. According to the concept of time in this book, **the substance of the universe wasn't created a certain number of decades ago.** There wouldn't have been time and the substance would have been still, if it weren't for chemistry and reaction among substances. But endless reactions and chemistry do occur among existing elements and materials; thereby, we sense time. Processes lead to results, like planets or galaxies. The substance in the universe wasn't created an endless number of years ago, because if that were so, then what came before that? And who brought the substance and where from? Again, the substance in the universe wasn't created, but just **is**, without time. Since there is a reaction, we sense time through processes. Our interpretation of the processes is in accordance with the sensations of the infinite "self."

Another question regarding our daily reality: if a car is driving 100 kilometers per hour toward the direction of the Earth's rotation, what speed is it driving at? And if we were to calculate the Earth's rotation at a 2,000 kilometers per hour speed, what would be the speed of the car then? And what is the car's speed if it drives in the opposite direction to the Earth's rotation? Having answered these three questions, what is the velocity of the vehicle, when in both cases we add up the Earth's rotation around the sun at, let's say, a 100,000 kilometers per hour speed? And what would be its speed in the scenario mentioned above if it is standing still? Again we reach the conclusion that time is an individual interpretation and the comprehension of the "self," which can sense time intervals and the fixture of situations in a subjective manner.

Let's take a different example the football. A ball is set on the ground. A fan watches the motionless ball, unaware of time passing by the still object. But from the starting whistle, the ball moves with the help of the players and the spectator feels a different sense of time than the previous one. As far as the ball is concerned, velocity isn't relevant to its stillness or its velocity of motion. For the ball, time stands still even when it is in motion. There are different laws of nature that apply to this ball, but this is a different matter. When the ball is kicked, G force is created (by the air's resistance) and the ball is drawn to the ground over Earth's gravity, this time without any G force. Only when the ball meets the ground, G force is created (a relatively massive one). Gravity doesn't have G force operating on the ball. Or rather – when a planet draws a certain object from afar, or an asteroid, for that matter, there won't be any G force operating on that object, since the Earth's gravity draws all mass toward it (of the asteroid or of the object).

In conclusion: if the ball keeps on moving toward a planet, from the spectator's point of view – it's moving at an enormous speed. But from a different man's point of view, a man who is in the ball, time in relation to speed is irrelevant. Meaning, the person inside the ball wouldn't feel any sort of change. Time is an individual interpretation of the infinite "self" and what he feels toward infinite situations.

There is a convention that speed is measured from the still object's perspective, but this is a convention set by man. For example, a man is running on a treadmill (conveyor belt) at a ten kilometers per hour speed and doesn't move ahead. The question is, what is his speed according to this man's set convention?

Again, based on the aforementioned example, there's a production of energy which is actually a process. Time regarding this process is an individual sensation made by an individual interpretation of the current situation, by the infinite "self." In fact, speed is the result of the production of energy...which is a process.

A final example : let's say we fire a rocket to the moon. The engines begin to operate at full speed, but the rocket doesn't soar, which means it doesn't have the time in relation to speed factors. But if we put that same energy in a lighter rocket, it will rise – but only from the spectator's point of view. Therefore, speed versus time is seen in our perspective as conscious beings. For the rocket itself, speed and time are irrelevant factors, unlike other laws of nature like: acceleration, heat, friction and their influence on the rocket.

The question is then: why has this law been concealed until now? Well, each individual has time sensations according to their interpretations. Their interpretations lead to an event that matches it. For example, an individual tells of an event that occurred two days ago, thinking it was only yesterday. As far as he's concerned, his feelings are legitimate. Yet if he was corrected and provided with proof that the event took place two days ago, then his feeling about the duration of the event would alter. Again, the sensation is legitimate, according to a new interpretation of the current situation.

There is no such thing as "cutting time short" or "folding space." only whatever exists in space may be twisted, but space itself is "nothing."

In an "absolute" conclusion, there never was, nor ever will be, an entity that has created the universe. The universe **is,** without time. The universe is "vibrant and alive" according to a process that takes place without the interference of the concept called "time." These are endless time sensations that flow in the body of each conscious being, without their awareness. Therefore, to us these sensations are legitimate. For that reason, we cannot discover what is legitimate and what matches our interpretation. That is the "Dark Maze."

This insight is supposed to change the confused consciousness of a man regarding creation and stabilize the chaos we are all in. Maybe this will bring us to salvation.

The speed of light. Theory
An astronaut begins moving away from Earth at the speed of light:
His spaceship is transparent.
The astronaut claps his hands in one second intervals.
There's an observer on planet Earth who keeps eye contact with the astronaut using a state of the art telescope.
Once the spaceship reaches the speed of light, the claps gradually begin to slow down for two seconds, as far as the observer is concerned.

When the spaceship slightly passes the speed of light (such a theory exists), the eye contact with the spaceship is disconnected and the image of the spaceship disappears from view. However, the astronaut's clapping continues at the same pace.

Time (which is a concept) doesn't influence substances. Time indicates the involvement of processes. Genetics and Old Age

Each type of animal has its circulation structure. I am referring to sex, and offspring, and the way of life. Animals are not reproduction machines. It's an exhausting process in its particular phases, accompanied by a sensation of "I have done my part."

We are born with a body and bone structure; a skin tone, eyes, and hair. It's genetics. **Genetics doesn't have an expiration date for the creation of a certain illness. Old age is not a genetic phenomenon.**

All that has happened during evolution has been embedded in genetics and passed on from parents to their offspring. In the past, an average woman delivered children by the time she was thirty (give or take) and not past that stage in her life. A man was fertile up until the age of fifty (give or take) and not beyond that. Therefore, there was no genetic continuity. Genetics was not involved past the stage of fertility of the woman and the sexual involvement of the man. In this case, how can we explain the fact that a man continued to age beyond the age at which he could no longer contribute reproductively? Genetics wasn't there to be exposed to intertwine with the aging process and it, therefore, doesn't carry an old age gene to pass on to future generations! For that reason: **the individual "self," which feels old, results in an old body.** This process is not symmetrical and doesn't occur throughout time. The body conducts itself and acts with an accuracy, according to the infinite sensations that recede and accumulate. This process doesn't occur depending on time or age (see **"Concept of Time"** chapter), but based on **sensations governed by interpretation of the events accumulating throughout life.**

Teeth, for example, can age at a very young age or, on the contrary, they can be in good, healthy shape at an old age. The digestive system, however, doesn't age at all. It runs by sensations of conduction (or rather lack thereof), regardless of age.

It has been said that the body's structure and operation equals the structure of the infinite "self" among all living creatures. The "self" is individual, of course. From the very birth of every living creature, it begins to accumulate events throughout life, so that at certain stages it carries a whole arsenal of events that match in sensations to an old, fatigued "self." The body acts accordingly. Since the brain doesn't differentiate imagination from reality, all that was in the past is currently vivid as far as the "self" is concerned. There is no mechanism in the brain that erases memories (good or bad). Because of that, the individual "self" starts to let go of its grasp of life, knowing that it and its sensations are living in the past. There is, therefore, a need to be rid of such sensations as: I created, I made, I experienced, I failed, I succeeded, or I grew tired. Those are "gray sensations." The body operates and acts according to the individual sensations of the infinite "self," for better or worse. These processes are legitimate as far as the infinite "self" is concerned, thus, there is no one to blame. Each living creature survives without awareness, assisted by diverse and endless strategies, supposedly in the "quicksand swamp." There is no logic or justice to this insight. Some children and animals are beaten and abused because their survival instinct exists but lacks grasp. This usually results in a quick drowning in the depths.

Theoretically speaking, a body without a "self" isn't supposed to age, since it is in constant renewal. Therefore, theoretically, **a body can live eternally**. As stated earlier, no living creature makes it to old age in a symmetrical manner like an hourglass. Rather, the sensation of time is individual as well. There are ups and downs, and overcoming hardships with the addition of the past. As a result: a feeling of degradation is created, and so old age comes with a worn-out, tired sensation. We should all be aware of this and continue living our lives in the **youngest** way possible. We mustn't count the years, but take care of their quality and focus less on the length (see concept of time).

One of the reasons for long life expectancy in our current time is the new culture that intrigues, interests, vibrates and "pulls" life forward.

There is no ticking clock toward the end. The end isn't always in sight, and life and the body don't have an expiration date. Time is nothing but a concept, and therefore cannot have an impact upon processes. For example, ice could melt in a few moments, over several years, after millions of years – or never. Time doesn't determine anything nor is it relevant.

In conclusion, old age isn't the same for all human beings. It depends on how old each separate individual "self" feels. There is old age that is the "shrunken self"; there is the old age of the "self that has lost its grip"; there's the old age of the "my brain is old, but my body is fine"; and there's the old age of "a generally old 'self." **The brain is old according to its owner's (the "self") sensation.**

The old body matches its owner (the "self"). The brain doesn't interpret. Theoretically, without the "self" the brain wouldn't deteriorate. But the brain does have an owner and that is the "self." When the owner is worn-out, the brain tires in response. All the theories regarding old age are ludicrous and irrelevant to the actual occurrences. And again, we should feel as if we are, as the saying goes, "riding the wave," not sinking below it. We shouldn't feel defeated.

We must learn to divide life into periods: to draw a veil over previous periods and start living all over again.

Animals in the wild, unlike humans, are exonerated from dealing and thinking about old age, about the end, about what lies beyond the end, about premenopausal and menopausal periods. These are other manifestations made by mankind, causing us to be helpless and anxious.

In the future it will be possible to expand the life of the biological body through wisecrack inventions, but that would be a tired and forced life indeed.

As for the phrases "old man" or "old woman," they carry a negative tone that might encourage violence. We should, therefore, learn to use the word "senior" instead of "old man," since it indicates more reverence.

IRONY

A sick child arriving at the hospital for treatment would most likely receive more attention from his parents. At the same time, this would only cause him to embrace his illness and enhance it without being aware of it. His mother will cry, buy him gifts and say: "Son, you have a severe illness." She will stroke his head with a concerned look on her face. She will hold his hand to relate to the tragedy. And this is yet another phase that would further enhance the disease. The authorized specialist will say: "Let's hope it will be okay." And that will also enhance his illness, knowing that the specialist doesn't cure illnesses as he should. Now that he is disconnected, the biological resistance in his body causes further retreat. When the boy's condition deteriorates, the authorized specialist announces that there is nothing left to do and only a miracle can save him. The specialist will then deliver the details of his illness, describing it to the patient's "self." From here on, the doors close on the child and his fate is sealed.

The course of operation in this case is therefore wrong. The principles of sickness and health are specified throughout this book. I recommend conducting workshops that will be attended by people who have entirely related to the principles of creation according to this book, and who also have knowledge of psychology, and together strive to change the "self" of the sick child (or any other patient of a grave illness) and transform it to a "self" that is above illness and that denies the former sickly "self." This is done to, so to speak, step out of one's current body and step into a new body, as one would do with a new suit (by a diverting sensual imagination).

One way of "diverting the imagination" would be to convince the patient's body that it is now beginning to cleanse, and setting a date for when it will be 100 percent illness free. When we speak to our bodies, we shouldn't use such phrases as: "You are sick" or "You are special," but refer to the person as "a survivor" and "a winner," thus changing the style of walking and the body language to a language that is aimed at a specific destination, beginning with the ability to walk continuously. We should allow our bodies to assume, for several minutes, the position of a tiger lurking after its prey: with knees bent, muscles coiled and tense, and without flinching. In this position, the body is frozen in place with fingers drawn; the facial expression is sharp, and the eyes focused and plotting. We should speak to our bodies daily, telling them in a convincing manner that there is an improvement in their condition (in the case of an illness, for example) and that the illness is receding. Theoretically speaking, complete disconnection between mother and child is an inseparable part of the creation process, but since it's not practical, psychologists and medical professionals should be encouraged to teach mothers how to make a 180° change in their reactions, so that instead of the pity that feeds the illness, they learn to convey pride in the child's bravery in fighting the disease. In the case of the sick child, this could only happen if the child truly and wholeheartedly wishes to let go of his illness. Because throughout their illnesses, most children feel privileged, loved, and pampered; so in a way, they actually learn to enjoy their own misery, seeing as they are visited by a lot of friends and receive many presents. In this state, a child is usually released from any responsibly, tasks, or work. So, theoretically speaking, if a child were to be instead shunned by his friends, who would label him "the sick child," it would motivate him to fight. However, since like the disconnection from the mother, this too is an inhumane and impractical solution, it is better to train the child to change his or her own body language into the body language of a fighter who is healthy and in control (don't enable others to mercifully stroke his neck or chin).

In addition, we should also learn to guide the sick child to climb up and down the stairs in the same way the preying tiger does: silent and alert. The child should practice moving very stealthily, with the guide occasionally making a rustling sound using a leaf or a tree branch to signal the child to stop. In this sort of game, the child becomes a tiger and whenever he hears the rustling sound, he ought to freeze like a tiger, with his muscles tightened and his back hunched; and after a few moments slowly begin to get back to the upright position. This should be done repeatedly. Of course, the guide should continue talking relentlessly in a half hypnotic tone, and repeat: "You are a tiger preying on a dangerous beast," so it will be more and more vivid. The guide shouldn't give explicit directions, but guide the child in present time, saying things like: "You are walking alert"; "You stop right away"; "You freeze in place"; "Your stomach is flat and muscular"; "You are aware of your body"; "All of your bodily organs are in their right place"; "Your internal organs are flexible and efficient, and your body is clean and pure"; "You feel this"; "You are aware of this"; "This is real and not the other way around." Do it once you finish the exercise, in order to teach it new habits. Don't flex your body while you sit or rest. Collect objects from the floor with a quality, bouncing crouch. Don't lean while you crouch with your hands on your knees, like when your abdomen and back are weakened and bent in a shape of a bow, and your shoulders are sloping. Teach the child to keep an alert body during the day. Let him be aware of his body that he is finally in touch with. And his body now has an owner. These workshops are suitable for everybody: a fat man, a skinny man, a sick man, and a healthy man. Obviously there's more. You should check on his mental state before, during, and after the training. You should construct a work plan that the child has thus far believed to be beyond his reach. You should remove blocks that have hindered the path to the ultimate goal to motivate the child to get it on his own, and return to functioning society with a sense of a "self" that is "above," and not on the fringes of society as a poor and beaten ego. Teach the child to return home, but not to the old maternal sayings of: "Eat. It will make you stronger." There should be no preferential treatment in the house of the healing patient and he shouldn't be handled with unnecessary caution.

That's what the tiger does. This is his reality. A tiger preying on dangerous prey equals a 100 percent efficient and healthy body. **But if the tiger had dysfunctional thoughts and behaviors, and shared the same beliefs as others, then it wouldn't have survived a single day.** You should relate to and imitate the tiger's "self."

As a final example on the matter, I present an example from a documentary program recently broadcast on television, which dealt with the case of a child who had gradually become paralyzed in his lower torso. Throughout the program, we witnessed what I could call "behavioral misses." That is, we saw his mother carrying him up the stairs and nurturing him endlessly, while the boy didn't cooperate, and didn't try to work with his hands to ease her burden. We could also see the child sensing his hands becoming weaker and weaker, as the weakness took over his lower torso. Still, he exhibited no "resistance," and the only reaction we saw from the parents was: "Oh my, the boy is sick. Careful! Don't move." An urgent call to an ambulance was then made. The child could hear every single account that was given of his condition. It was a tragic atmosphere. The seemingly authoritative specialists confirmed his condition in his presence, so then he was convinced of his illness and it thus became chronic, according to the sick "self."

The internal sacred order calls upon all specialists in the field of psychology and hypnosis: just as it is possible to paralyze a hypnotized person for the sake of trickery and showmanship, I am certain that you can discover ways of using these principles, in a positive way, better late than never.

MENTAL ILLNESSES

Each individual lives in an individual universe: his own private world (that's why it's hard to understand the other). In some situation the individual is caught in confusion and embarrassment within the haziness of the maze. As we know, the maze doesn't contain fixed signs indicating how to exit, so the individual is trapped and helpless in a dead-end, where he gets lost, according to his deteriorating interpretation. Proper treatment, or on the contrary, aggressive treatment of this poor wretch, would only serve to further complicate his situation in the maze, and there is no magic potion that could get him out either. Since we by now accept that the body and the brain operate by the "self," according to its sensations and individual interpretations, then the individual should be guided by logic and reasonable phrases.

We are all born with a brain that has wasted abilities, because we do not carry out "directions of operation." Therefore, each person uses it according to their "self," and so all paths and beliefs become legitimate as far as the individual is concerned, and that is the human tragedy. It is therefore important to continue incorporating psychology lessons in schools, for if there is no awareness, the human tragedy will grow. A child will grow up with no "steering wheel" to hold on to, or rather driving while holding the "wheel" the wrong way.

The Mental Illness Principle

Schizophrenia, for example, isn't a cerebral phenomenon. It occurs when the individual "self" doesn't have an authoritative personality designer. A person flows wherever his hallucinations and imaginary thoughts lead. He embraces them without control. Because of that, every character he possesses he finds legitimate and his behavior becomes the same as the persona he has identified with, based on the interpretation of his "self." Had he been aware of his situation and of its principles, he would have driven the wheel to the center of the road and back to "reality." Since the brain lacks comprehension and an imagination guide, it is the one to blame. When a mental patient diverts from reality, it is because his "self" is leaderless, and this needs to happen for him to come back, and for his "self" to authoritatively order his (non-understanding) self to return to the "center of the road" immediately. Some people program their minds and use them in the worst possible way.

When the reality of life slaps a man across the face, he is in a fragile state that might cause him to drift away from reality. This can occur if, let's say, someone believes in the existence of supernatural phenomena with scary names like "possession" and "demons." Such a person would surely imagine them. In this fearful state of the upcoming scary and threatening situation, he would feed his fear by acting out his imagination, believing that "What I feared happened to me." The situation becomes more real and tangible. That person consequentially falls into that same trap that he had feared and that the "self" had created for itself. A man can go crazy in such a situation. Again, don't blame the brain. It is the "self" leading itself into dark places.

With schizophrenics, one identity can sometimes lead to another. This happens when the schizophrenic is passively swept away by his imaginary thoughts into unknown and unexpected places. He is a "bad driver" and cannot direct himself back to the center of the road.

Twenty tons of serotonin couldn't cure these types of phenomena.

On one of the National Geographic channel programs, a guy was documented after claiming he saw ghosts pulling on his leg. His leg did move during his sleep. He wasn't lying. Hidden cameras were then installed in his bedroom to follow him and see whether there were actual ghosts in his room. As expected, nothing could be found. We are in the twenty first century and people are still trying to track down ghosts. That's how it works: the "self" believes that there are actual ghosts. The "self" is scared of them. This fear of ghosts fuels the imagination. Since the body is operated by the "self," the leg moves when a situation in which ghosts are supposedly pulling his leg becomes vivid in the imagination. This is a legitimate situation, as far as the person is concerned. But ghosts aren't the ones moving his leg. That is the determining "self," convinced that ghosts are present in his room and pulling his leg, (according to his sensations, which are preceded by his individual interpretation). And that's how it shall be. I have pointed out throughout this book more than once that the brain doesn't differentiate reality from imagination or a dream. The brain is obedient to the "self." The brain is the connecting extension between the "*Ultimate I am my own maker*" and the body. So you can sleep peacefully; ghosts only exist in imagination. **The brain is not ill. This phenomenon isn't a brain dysfunction.** So don't disturb the blameless brain with various potions and remedies.

Since we've established that the same principle that is true for humans applies to animals as well, we could assume that whatever the individual animal thinks or believes becomes real, as far as it is concerned. And the animal's body will react accordingly. Let's go back to Flash Carbonado, for example. He was extremely mentally ill. He experienced his first crisis when the entire herd in which he grew up was sold, along with his mother, and he remained alone. Any foal in his condition would experience shock and mental stress. To him, it was mental rape and an unthinkable misery. Since sensations such as these don't disappear from the memory, he would most likely carry them with him for years to come and might also develop, in the best case, a fear of abandonment, with a chance of a severe deterioration in the worst case, according to the snowball principle.

In addition, since he had gone through all those tribulations, even when he was finally back with me, at a much later stage of the story (see, "**My Story**"), he still continued to get worse. It was as though he had exited reality and given up all hopes of his life as it had been before the crisis. Gradually he became paranoid, and every time I would transfer a mare from one booth to another, he would begin to panic, fearing I would take her from him. Each sound he heard in the distance, he would interpret as the steps of a mare galloping away and deserting him, and that would result in his panic-stricken cries and merciful weeping. He went through the roof with hysteria and tantrums, suffered diarrhea and sweat spells as if he were begging for mercy. Moreover, I had to cope with serious threats to his wellbeing; it had gotten to the point that even distant sounds of a rooster calling or a cat meowing would be perceived by him as the sounds of a mare leaving, despite the fact that all the mares in the stable were standing right next to him. I had no choice but to take him out, lead him in the direction of the sounds that he had heard in the distance, and try to calm him down. As we could see, his body reacted according to his paranoid sensations (of, dissolution, and mind and body distortion).

So animals are just like us. You could even say that their souls are even more fragile than the human soul, since they don't have a language to express themselves and reveal what is in their hearts. They cry out for life just the same as we do. They have different moods like we do: submission, depression, hate, jealousy, fear, shock, anxiety, boredom, excitement, despair, love, mental illnesses, the love of creating, and aversion to creating. The horse either runs for his life or attacks in defense; he might also become passive, humiliated and submissive, exhausted and lacking energy. As for the neighs that he makes in despair, no one neigh is similar to another. Each neigh expresses a different sensation: request, pleading, complaint, disappointment, fear, despair, aversion, satisfaction, need to join a crowd, crying, hunger and thirst. That is an abridged language created out of the expression of endless, instinctive reactions.

As for paranoia – the paranoid is in fact the paranoid "self," convinced that someone is chasing it. It could be real or imaginary (without the brain differentiating one from the other). Both cases are legitimate as far as it is concerned. The imagination begins at a certain point, like fear, and then runs relentlessly according to its deteriorated and guideless interpretation. As you may know, threat and persecution both exist in nature, but on the other hand there is also the freedom to run.

A captured animal or a man who feels hunted is still trapped, and therefore feels (according to his interpretation) that the attackers signaled him out as an exclusive target, trapping him from all directions with no way of escaping. Neither man nor animal can, in this situation, predict the time or place of the attack, or the attacker's identity. Moreover, both are convinced that they cannot defend themselves against all this, so all they can do is imagine themselves living out the terrifying scheme. It is a feeling of constant horror, of an extremely evil paranoia. Serotonin and other substances aren't to blame for these phenomena, nor should the brain be blamed. It's an individual figment of the imagination, focused on an original fear, like the fear of cockroaches.

A closing message: the illusion of magic solutions should be eliminated at once. Serotonin isn't the cause of these phenomena, nor is any other substance. The level of these substances in the body equals the frantic sensations of the owner of the body – that is, the "self." We should, therefore, stop probing the innocent brain with various CT scans and invasive treatments (like electric shock therapy), and refrain from driving a miserable person crazy, reporting that his mind "was lost." Relying on the principles of this book at an early stage in school will save you a lot of misery. As for electric shock treatment, for that matter, a slap that brings about a recovery can also help.

Hunting

Hunting – that's what the "human creature" decided to call it. Better yet, you should say: **murder!** Yes, this is murder!

I want to share with you a scene from yet another movie, in which a beautiful actress aims a rifle at an animal and shoots it. She bursts out wildly laughing, filled with happiness as the animal loses its soul. You should know that she and the "brave" hunting instructor keep a safe distance from the animal. Where is the bravery then?

Why would you shoot an animal at a safe distance and think of it as a noble or fun act? The first steps to changing our perspective will be complete, once we acknowledge human nature. This is so that we may understand what should be corrected and where we should begin.

There are farms in Africa existing for the sole purpose of raising animals and then releasing them for hunting!

These hunters condition the payment on the fact that the animal will be freed from its habitat and running for its life. The more dramatic the act, the more challenges the hunter faces and he pays more to the animal breeders. Each animal has a price.

I should point out that the satisfaction those miserable beings gain from murder is massive, full of happiness and excitement. It manifests in hugs and kisses after a successful shoot, and compliments like: "What a beautiful beast!" (once it's already lifeless).

There are those who set fire to domesticated animals while they are alive out of sheer boredom, a grudge, hurt, and twisted pride; and the authorities remain silent. Some cults set fire to domesticated animals while they are alive, as victims in the appeasement of some "gods." There are restaurants in India that have on their menus a dish of eggs cooked with the chick still inside them. No one thought, not even the tourists, of the process in which the chicks are cooked while they are still alive.

As long as this is human nature and conduct, there will be no concrete reason for our continued existence.

Everybody knows that elephants and rhinos are hunted for tusks that supposedly improve the libido of the inferior, degraded, believing, and idiotic man; particularly modern man, with the means to purchase this remedy. So far it's all known. Yet what about those tiny elephants in Asia, deprived of their mothers and their families? Can you comprehend the extent of the tragedy involved? Can you imagine the brutal and tragic methods with which the so-called civilized abductors "train" the poor baby elephants? (I will save you the horrific description). The baby elephant experiences awful horrors. The atrocities of the massacre to his family and his mother accompany him his entire life. This is the place to let tourists know: while riding an elephant, you are becoming an accomplice to these dark atrocities, without even being aware of it.

These sorts of actions are mostly followed by horrors that don't reach the media. Like when the hunters tear the tusks from the mothers' carcasses and the miserable orphan is horrified. He remains close to his mother's carcass, in misery and agony for days, up until his demise.

I know we don't all do that, but we are all in the same sinking ship. Most of us don't care. Eventually, almost nothing is being done. There are states (in Africa, for instance) where almost all animals have been exterminated. Save Africa! Where are the leaders and legislators?

What a shame that we don't have a creating, operating, and accountable entity.

Poverty drives men to hunt, but so does wealth. By that logic, the question that should be asked is: where are the fine qualities of the human being? The human has an urge to trample more and more. It should be said out loud. When nothing remains to be trampled on, he turns to weaker beings. Immediately killing a bull in a bull fight isn't challenging enough. It's not exciting anymore. Excitement is achieved only by causing this bull grievance, humiliation, and misery; by stabbing knives in its body until the defeat will be completed with its demise. Only then will the man be victorious, according to his twisted interpretation.

Orangutans, elephants, tigers, and a large variety of animals in the Borneo forests will be extinct during the next several years due to the elimination of these forests at a ruthless pace for needs of agriculture and hunting. There's no way out, since man must finance himself in order to survive and a war of survival is a legitimate claim.

A moment of cynicism: after in-depth research that I have conducted with the finest mathematicians, biologists, and other scientists, using state of the art microscopes and highly complicated mathematical formulas, there's only one conclusion to be drawn: **limiting childbirth**. Yes, limiting childbirth and maintaining the equilibrium has been done by animals for millions of years, using their tiny brain in comparison to the super brained "smart" man. Among other things, limiting childbirth will solve housing problems (and will also decrease uninhibited, massive construction). If mankind cared, then perhaps we could boycott chemicals used in agricultural and other products that destroy the forests.

Hunting's effect on youth

Young people are confused because they receive mixed

signals. They are taught that they shouldn't harm other living creatures, but not only is the media drenched with violence, there are actually television shows dedicated to teaching them how to hunt (like "Sunday Hunting"). How then will youngsters behave once they grow up?

All the efforts to sustain the world won't help if childbirth isn't restricted from this moment on, because in case you haven't noticed, it's already too late now. Another example that demonstrates these most problematic phenomena is of a poor woman with fourteen children; the youngest – only one year old – and she is about to give birth to another. She cries out and protests her financial state, her shortage of food, and the hardships of raising her children.

My message

We should use our non-comprehending and wasted brain in a more efficient manner, without robotically embracing fixed norms. Let's take some responsibility for our actions, the sooner the better, because man is riding toward the abyss.

I would like to take this opportunity to address the states that permit whale slaughter on their territory: save the whales! Whale slaughter is a crime. The criminal hunters dissect the agonizing whale's body, while it is still alive and breathing, begging for its life with a look of horror in its eyes. If man doesn't become conscious of his actions, they will return to slap him across the face.

A message to all of you who still have a head on your shoulders: eating whales' meat is like being an accomplice to a crime!

Also, you should immediately stop using remedies that originate from animals. These potions and other remedies don't help. These are miserable beliefs that only spread in human society. Their use is common only among naïve individuals, who believe anything that's distributed, without questioning or objecting.

As for orangutans, hunters beat the mothers and the rest of the family with arrows and axes. They grab the babies and sell them to the highest bidder. The same babies are kept in a tiny cage for the rest of their traumatic lives by the scumbags…And the world remains silent.

I would also like to take this opportunity to address nature resorts' supervisors – you should mind what is going on around, because nature is not something you should take for granted. In addition, you should rescue agonizing animals, or humanely put them to a fast death.

AUTISM

If life is indeed like a "quicksand swamp" and it is not to be taken lightly, then what could be easier than not fighting the war for survival? Indeed, our brain isn't used automatically – it is the "*infinite* self" that is supposed to operate it. There's no particular entity that makes sure we are born healthy and in good quality or vice versa. We are born into a war for survival. Since a child is born without instructions in the brain, the beginning of his growth will be critical. That's because each person subconsciously chooses his own strategy on how to grasp onto life. He is born in a living and breathing body, without him choosing so. Similarly, the animal in the wild learns the laws of survival also without being given different alternatives to choose from. The one difference is that man still has a wide range of options.

The autistic person chooses, subconsciously, to remain hidden in a faraway corner from the quicksand reality, where he hides in a type of bubble, knowing that it's his home. Everything outside the bubble is a scary and quicksand world that belongs to others. As a result, the more he is treated as a different and problematic child, the more his home – the bubble – becomes dominant, real, and tangible. Therefore, we shouldn't treat the autistic child like an imbecile, but as an adult who owns his life. We must remember to consult with him on various matters, but not like a parent treats a sick and privileged patient, because that way he'll remain sick and privileged. But more like a friend talking to an understanding friend. That way, the child will have more courage to step out of his bubble, and enter the threatening and quicksand society. It's a borderline and diverse situation, but the possibility is out there.

An example of this would be the schizophrenic "self" that assumes the role of a woman and then acts like one. And if it goes on to assume the role of a big movie star, it would then also act accordingly. When the individual returns to the "self" that had shaped and raised him to be like this, it would determine his behavior. The brain doesn't hold any responsibility or an ability to interfere. It is the infinite "self."

That's why we hear of amazing mental abilities among certain autistic people. The brain is there and it is not damaged. But sometimes, when a certain autistic person overcomes the hardships, the threats and chores outside of his bubble, his world is reduced to focusing on one field that gives him a more mental energy to draw information from the subconscious to the conscious level. On the contrary, another person might think abstractly and infinitely. It resembles a hypnotist, who takes the hypnotized person out of the quicksand reality by getting him to focus on a particular field, in a more efficient manner than of his abstract reality. That is, the hypnotized man is willing to let the hypnotist "drive" his "self," which means that the use of the brain is infinite.

I must stress again that we shouldn't treat the autistic child as though he's different and an outcast, because then he will retreat to his home (his bubble) in a faraway corner of reality. We should instead, treat him as if, "You are a comprehending individual who makes decisions. The world is yours as well. We are counting on you"; whether that means letting him run some simple errands or try to find a solution to his own problem, or even a tougher task. When the time comes for the autistic child to take his first step out of his bubble, and into the cold and challenging reality, there are two possible reactions that he might have: one is that he will return to his bubble, because the outside world is too threatening for him. The other is that he will gather the courage to step out of his bubble and into a reality where the skies are the limit. It's like diving head first into cold and unpleasant waters.

On one of the "National Geographic" programs, there

was a woman, who had been autistic, but is now writing bestsellers. One might say that this woman dared to venture out of the autistic "self," at a certain point, once she felt a firm grip outside of her autistic bubble. I am convinced that she looks back occasionally at that girl, the autistic one, and maybe feels a sense of missing out on her youth.

Comment: this only concerns situations where there are no birth defects or flaws.

Theoretically speaking, even if the brain were larger, these principles would remain the same. The brain size is separate from the infinite "self," which operates it. The individual trains others around him without being aware of it, for better or worse. This is also the case with horses: I had a mare that would only walk backwards. It wasn't any sort of cerebral illness or blindness that caused her to be oblivious to what she was doing. She was simply frightened of her owners and reacted by constantly retreating backwards. Her frustrated owner was mad and responded violently. This obviously snowballed since the mare only retreated farther and became even more fearful, while her frustrated owner carried on with his blind and indirect training. From then on, the mare's strange walk became fixed forever.

Epilepsy

During an epileptic episode the patient loses consciousness for a moment. As in the previous chapter, this too is not a brain malfunction, but a phenomenon that only arises when a morbid feeling is awakened within the patient that leads him to feel as though there is a void in time, which then leads to dizziness and passive disconnection.

Once the individual experiences this sort of phenomenon, he might acknowledge it and embrace it for years to come. As far as he's concerned, he is verifying every situation of this sort to come, by a feeling that says, "My worst fears will be realized." Then thoughts turn into actions.

In nature, this situation is ironic. It's true that the tigers, for instance, are frightening, ruthless predators. But ironically, those tigers maintain the physical health of all those around them by forcing them to be constantly alert. Therefore this sensation of dizziness doesn't exist in nature, and neither do passivity and deterioration. Those are the real laws of nature.

Recommendations

A theoretical example: say an epileptic patient finds himself alone in the jungle, with threatening predators breathing down his neck, in an all out war for survival and existence. The question that arises is: will that patient suffer an epileptic fit in such a situation? The patient will operate on his "resistance" ability, claiming that "This is where I don't give the phenomenon a chance, because I decide upon the course of my body. Each time this emptiness occurs over time, I instinctively pull myself together and turn the situation into one of a 'survivor in a wasp's nest.'" In this situation, the fingernail grabbing instinct in life manifests for the first time.

URGE AND LACK OF URGE

Physical and mental efficiency are better felt when sensations or urges to perform arise. No human or animal should be forced to perform an action they have no urge to perform. It cannot be done by giving them various potions or by using force. One can't love or hate by will, perform a sexual act without sexual attraction, or run without an urge. Allow me to paraphrase – it *is* possible to perform these acts, but there wouldn't be a real sensation of urge behind them and they would then be ineffective, like faking a laugh, faking a cry, faking an orgasm, and so on. That is the reason why one cannot buy "will power" in the supermarket. You can affect the alteration, from lacking the urge to a state of having an urge, by using reasonable tactics like giving proper compensation upon completing a task, changing strategies and so on.

Our efficiency is expressed due to urges and desires or, on the contrary, it isn't expressed at all due to lack of urges and desires. Let's take a boy, for example, who is too shy to use public transportation or ask for rides from his neighbors and friends, for unconscious reasons like laziness, shyness, etc. If he falls in love with a girl who lives far away, his love for her will drive him to use any transportation and even to get to her house despite the obstacles standing in his way (which would suddenly appear less substantial). Difficulty is therefore not a physical element, but rather a mental one. It is governed by the "self." And so it may be said that **we are efficient only when there's an urge carrying us to fulfill it: once there's a possibility to achieve it.**

The story of the boy in love is an example of conduct. Meaning, if the girl he likes doesn't return his affections and rejects him, his "self" will experience a kind of "short circuit" of sensations. It's a "short circuit" that occurs between the conduct and the clashing emotional humiliation.

Many people find it difficult to maintain a healthy lifestyle. The same is true for horses training for a race: if the horse doesn't like to train and doesn't wait for the saddle to be bridled, it won't get in shape even if the trainer rides the horse and works with it every day, against its will. The only thing that would happen is that the horse would suffer various aches and pains, such as back aches, joint aches, and cramps.

To train the horse and make it run, the horse must be in an excellent mental state. It must love its trainer, the training itself, the compensation following the training, the jockey (who rides it), the stables, the stable's other inhabitants, the entire atmosphere, and the food. **In short, everything that that place and life in general represent to it. The same principle applies to human beings.**

Because of that, since we (including animals) aren't biological machines, we must use our brains in order to think how we can make a horse love and do what it does, by a sense of **urge** and not out of obligation. The methods should be personally adapted to the horse's individual preferences. One option is to train it along with another horse, at least to create motivation when one pushes the other up the hill, encouraging him.

Without impulse, a horse can train or run by itself, but it will also breathe heavily and sweat, without actually becoming physically efficient. Over the course of time it will deteriorate. There begins the "magic cycle" (or the drug cycle) in which the horse will no longer be efficient. At that point, most trainers would administer vitamins to the horse, along with protein and various shots; some will have the horse checked by different vets and make other attempts that will be doomed in advance, as the horse will obviously continue deteriorating. The horse might duly carry itself with a sore side and loose joints, ulcers, damaged kidneys, and whatnot. But then it would only be considered by his trainer as a second rate horse.

To sum up, physical fitness has to do with the sensations of the "self" rather than physical training.

Animals in nature don't practice; they are loaded with a constant sense of fitness. It's a sense that says: "The predators are pushing me and forcing me to be alert, on guard, and fit at all times."

Another example of changing from a state of having no urge to a state of experiencing the urge: a guy is in love with a girl who doesn't return his affection. It is likely that by changing his courting methods and adjusting them to the object of his desire, the girl will change her position from lacking urge to a sense of urge. This is an example of a way to change feelings by changing your interpretation of that situation, i.e., by simply changing strategies.

Without urge one cannot cross the ocean, especially not during this era of development. Without urge one cannot cross the desert, neither by camel nor donkey. An individual who is loaded with urges can sense the future situation as one he has already put behind him. In fact, he senses the journey and its end, with no obstacles or fears.

Exercising without any urge or inclination to do so doesn't contribute to our health or our physical fitness. Sometimes it even generates a health hazard. Take a man who starts to exercise due to social pressure, yet he carries certain emotional baggage from the past or present, and perhaps even fear of the future. This emotional strain might cause him to sense a certain flinch in the corner of his heart or even develop other health issues. This person wouldn't feel an improvement in his physical fitness and might even be harmed. That is because he has no urge to engage in exercise. On the same scale, there is a possibility that by changing the situation from dull and unchallenging exercise into rhythmic music **that he likes**, then this person might sense an improvement in his physical state, as well as in his health and mental state.

I hope I am not encouraging lazy people to be even lazier on account of this insight. Please refrain from using my statements as a "get out of jail" free card against exercising or as an excuse to go on being feeble. We mustn't be proud of our illnesses, or use them to gain society's sympathy, or even benefit from them subconsciously. Flaunting one's disease or misery indicates an internal label that says, "I am sick." There's no way out of there. This only causes the illness to become chronic, as it deteriorates to the dark abyss. We should immediately stop conducting such casual chitchat like, "I have been diagnosed with this and that germ," or "I have this and this issue" or "My situation is more difficult and I can't do this and that..." People like that don't have a place in nature and wouldn't survive in it a single day.

HAPPINESS

Laughter is a positive, yet temporary, sensation. I should add that laughter, like any other occurrence, originates from within infinite sensations. Meaning, there is laughter through tears, there's mocking laughter, there's laughter that emerges from moments of happiness, and there's laughter that derives from a mixture of sensations.

Happiness is accompanied by a continuous feeling of harmony, like a bird soaring to the heavens. So does the physical sensation within the metabolism. Under such conditions, the function of the digestive system will be good, and conduction will occur regularly. Not only that, but cognition will improve, the blood will improve, and the immune system will become stronger. The appearance of fungi and inflammations will be less and less, all in accordance with the varying sensations. Yet it is a given that we can't choose between happiness and sadness, so it's not the wisest thing to recommend simply being happy. Yet the option does indeed exist because, as we have by now come to realize, it *is* in fact possible to change a negative situation into a positive one by changing one's interpretation of the same situation. For instance, if we owe a hundred thousand dollars at the bank, it's recommended to think, "I'm lucky it's not a million dollars," to learn a lesson, and try and make it right the next time. We should also sustain a positive body language, because society will react to our demeanor. We are in an existential war to be "more" than the other person, without even being aware of it. The question still remains: don't happy people get sick? Well, obviously there is no specific entity saying: "Be happy and I'll take care of your health." Happiness, as you may know, was the first condition to life. Because otherwise, why live in the first place?

In conclusion, the sensation of happiness doesn't absolutely guarantee good health. We should be in sync with our bodies and manage them by their rightful owner the "self" (explanations provided throughout the book).

ALLERGY

Allergy is caused by anything that the "self" is allergic to, in an individual way. The examples are numerous: certain substances, smells, foods, a certain taste, dust, a certain powder, a perfume, the smell of mold, and more. All these are causing the "self" to develop an individual and intolerable sensation by interpreting them as an allergy, in addition to the physical allergy. Keep in mind that all these things can occur by memory and imagination, not necessarily by physical exposure.

The manifestations are also individual: red burning eyes, a rash through the skin, a sense of itching, or a runny nose.

Again, don't blame the body. It is the infinite "self." Since we are not machines, each of us has a liking, an aversion to, or repulsion by a certain taste or smell that the "self" is allergic to, or even by a particular atmosphere. Usually, the allergy is created when a certain smell or taste reminds us of a specific unpleasant situation, such as a traumatic event from the past that we weren't aware of, or are aware of at times. May I remind you that focusing on the past is, in fact, the present (see "**Concept of Time**" episode). It is a growing inclination among sensitive and emotional individuals, who live in a closed and sterile "nursery," and who don't see themselves as a part of nature. A perfect example of this would be the one I presented in previous chapters – that of a man holding on to a limitation as a playing card, and he's sure that he ought to be treated as a privileged being, and be given less tasks and responsibilities. Sometimes the events that lead to an allergy are a combination of circumstances. Obviously it all comes down to the present situation of the "self."

For instance, being lactose intolerant is caused by rejecting milk, according to individual tastes and smells. Once the "self" feels uncomfortable with this white beverage, sensing an aftertaste and an after smell, the "self" processes the unwanted substance in the body based on its own sensation. Theoretically, if we disconnected the "self" from the body, then the body wouldn't have any limitations, barriers, hardships, or illnesses. The conduction in the digestive system would be even more effective and faster.

But for that to happen we need someone in charge of the brain and the body.

If it's feelings we are discussing, here is an example of a sensation we might have missed: we would prefer a cold watermelon over a warm watermelon despite the fact that there is no difference regarding the taste (by the biological sense of taste). Therefore, the preference has nothing to do with the sense of flavor, but rather individual sensations. A different example would be cold soup from the fridge, or any other beverage that the "self" considers to be cold, but is actually warm and vice versa. If we were biological machines (that is, if we hadn't had the "self"), then the body wouldn't have had the demands and preferences of its sensations.

In conclusion, the digestive system does in fact function, though not independently. The food is processed by the "self," based on past, present, and future experiences; flavors and smells; personal taste; mental conditions and so on. All that I have mentioned is interpreted into individual sensations, according to which the digestive system operates, along with the rest of the body.

An allergy to a bee sting

There is no sense in the fact that even a minor portion of venom can affect and wreak havoc on the body. Still, this can happen even to a horse weighing half a ton, and yet again, it is the "self" that develops an allergy to that particular sting by the principle of the deteriorating chain of reactions.

Skin rash

One must "resist" the blossoming of plants and flowers, or any other allergic phenomena (see "**Resistance**" episode). That is, divert the situation and the direct appearance to a healthy skin.

In one karate method, the person attacked takes advantage of his movement, or of the attacker's fist, to turn it against him. You can imitate the same principle. In other words (regarding a blossom allergy), say to yourself: "I need the smell of the plants blossoming. I can't do without it. For me the sprouting is like a breath of fresh air. The blossom charges me with energies. For me it's like fuel and oxygen."

An important general insight regarding various illnesses

One of the most important insights is that in nature, there's a better sense of conduction, since the situations never repeat themselves. There is no routine. Each moment is new and different from its predecessors. The threatening situation in the wild, like the predators and deadly battles (for food, women, and territory) push us, with little choice, toward the sense of conduction. Though there are tragedies and loss, in a short while the situations change. The focus upon the new and unpredictable situations causes the dissolution of the previous tragedy.

Unlike life in the wild, man lives in a cyclical and tightened framework. He's focused on recurring and dizzying sensations. This way of life leads to mental and physical hardships altogether.

As for the spinning and dizzying sensation: it's the "self" that feels dizzy from the spinning situation. You can picture it like a man, standing in a sealed cell on a spinning merry-go-round (the image of a spinning merry-go-round might cause a dizzy spell).

A STROKE

A stroke occurs when a man turns passive and submissive to his hardships. This puts him in a state of a retreated life, slapping him across the face and controlling him. He can see it flash before his eyes, as he focuses on it and lets it push him around without fighting back (see "**Resistance**" episode). Even when routine accumulates and makes you sick, at a certain point the passively transformed individual lets go of his grasp on life. All these options are a mixture of circumstances. Since the body and the brain are managed only by the "self," the brain collapses according to the receding and submissive sensations of its owner. This is similar to nose bleeding, when an individual feels exhaustion, fever, dryness, mass collapse, and a breakdown.

The treatments administered on patients who suffered a stroke are not enough, since these people are treated as "diseased."

Prevention and Healing

We ought to learn to stare reality in the face, and this time resist it, push it away, and conquer it by "shoving our way" through it. That is, the patient should keep his head high and not think of himself as "ill." After all, when the "self" thinks and feels "ill," the body and the brain become ill too. It is recommended to participate in classes that require the utmost use of the mind and body (like tennis). When a person is governed by an urge to maximize the benefit of their body, the body operates in response. Making plans for a new life, different from the previous one, like moving out, changing the atmosphere, or even breaking up with your partner and moving on to a partner who is more suitable to our individual character (not before being advised by an authorized individual). The aim is to climb up, with our heads held high, and reach the point of conduction, without disturbances, but rather with a new atmosphere and rejuvenated energy. Most importantly, we should remove ourselves from any "suffocating" surroundings.

In order to fully comprehend, one must read this book in its entirety. Contrary to the claim that the present situation can't be changed, I say that it's enough to make a distinct change from the previous condition to the new one. It all depends upon the individual. That is, if he didn't have an urge for a new plan, it wouldn't quite work. That's because if the internal sensation is still "ill," like it was prior to the event, and the subconscious situation hasn't changed, he will continue to get sick. On the other hand, if he is charged with new energies, with a spark that pushes him to pick up life again, the brain will react according to the new sensations of the "*I am my own maker*" and will be adjusted by the new "self" depending on the new, brighter, sensation.

I should once again mention a difficulty: The will to live does not count. There is no one to listen to our wants. The quality of life is determined by the volume of sensations rising high or alternatively, by a sense of utter collapse.

LEARNING DISABILITIES

It is acceptable to think that a child is naturally prone to learning, due to his big brain. But we should consider the fact that childhood, along with its legitimate energy and challenge, collides with the schooling framework, regardless of the brain's size and ability.

I have already equated the brain to a super computer. Say you bought your child a super computer but he wouldn't be that excited about it nor challenged to use it. Is it a problem with the computer? The same is true for the brain: It is there, but the child has no urge or challenging desire to learn how to use the brain for a purpose that isn't his at the moment, and which he doesn't like. As far as he's concerned, the situation is legitimate and he doesn't consider it to be a problem. The problem exists with those who demand and expect things from him: to love that which he does not love. That's not all. Because in school, all we can see is the results that the child achieves. No one bothers to look for the reason. But if we look for it thoroughly, we would typically find it within the house or in society. Honor, respect, appreciation, encouragement and rewarding might have a good influence on the child. Again, that is the infinite "self." The brain is there and it isn't damaged.

Let me tell you a story about a child who, as far as anyone could see, was treated kindly and lovingly by his parents outside the house. However, in the house, the child was seen as a spineless, good-for-nothing. Understand this: **the individual child is, first and foremost, unconsciously overwhelmed by infinite difficulties, such as: coping, adjusting, assimilating, growing desires, laws, rules, conventions, chores, prohibitions, set frameworks and so on. Therefore, we shouldn't expect any magic solutions to solve the child's internal difficulties. A child is not a brain. Each child has his own "self," and it must be empowered and balanced as much as possible.**

In conclusion, as life is challenging (in the positive sense), so should learning be (see **"Urge and Lack of Urge"** episode).

ARTHRITIS

Arthritis matches the sensations of the "self." When the individual "self" feels dissolved and disconnected from life; we begin sensing disharmony or feeling as though we are losing our grasp; or when our spirits are low in general. Obviously the reasons are individual and depend on external sources rather than biological ones. It can begin during seemingly insignificant social events, and develop to an enormous phenomenon, leading the general condition to an abyss. The individual reasons are varied: endless accusations, sarcastic comments, "poisoned arrows" being pointed at him, a bruised and reduced ego, a demeaning attitude that dissolves the sliding and powerless "self" lacking in energy. The "self" dissolves and slips downwards without a firm grip. The body is created and operated only by the infinite "self." The "self" feels deteriorated and rusty and so does the body, with no relation to physical activity. There are no worn-out bearings in the body caused by mileage or lack of oil. Each person's sensation is individual, according to his worn-out and individual experiences.

Sometimes it happens that the "self" is completely worn-out, even if the individual is properly treated. There are German Shepherd dogs that receive good and loving treatment, and yet they suffer from joint pains. It's a state in which the dog's "self" is captive within an obsessive urge to please its owners, uncontrollably. The more the owner is happy, the more the dog is at ease.

Possibility of rehabilitation

There *is* a possibility for rehabilitation: start living a new and healthy life, physically and mentally, and try understanding the other. That means: conveying the body language of a proud rooster, with its chest swelled up with pride and its head held high; thus, projecting to the world that "I am in control of my life."

Society will treat you according to your body language.

In addition, you should make drastic changes in life, in areas that were up until that point shut off from you. Also, join different workshops or groups to break your daily routine, and encourage the downgraded and rusty "self" (like laugher workshops or dancing). You should maintain variety with different workshops. You shouldn't be hooked on just one particular circle.

FLOATING MEMORY

I am reminded of a disturbing event I witnessed when I was four years old: my neighbors were raising about twenty turkeys in their backyard. Back in those days, they treated turkeys just like food, and not as living and breathing creatures. There was a wounded, limping, and rejected growing chick inside the chicken coop. The adults picked at him with a stick until he bled. This image returns to my mind today because of my awareness. I can remember in detail how the poor chick fought for its life day after day, making horrible squeaks and crying miserably, in reaction to the hurtful poking. Added to that is the image of him barely dragging himself with a swollen and wounded leg. He was pushed to the corner, desperately attempting to step out of the terrifying situation, up to his coming demise. My heart goes out to him. I wonder why they didn't take care of his well-being, and remove him from the alienating and violent company.

During one of those dark and ignorant days we grew up in, I caught a small rabbit in the field and brought it home. Our dog immediately approached me and began barking at it. The rabbit, for some reason, fluttered and died. I can see now that its heart had ossified, fell silent, and it died after suffering a heart attack from the fear it had felt, being held in my hands with no ability to instinctively express its reaction and flight (though it yearns and struggles for life). Yes, they are desperate and eager for life just like we are.

I shall add as a note that rabbits, among other animals, have been almost entirely eradicated from the area where I grew up.

Another story I witnessed as a child:

It's a tale of a simple man, with a carriage and a horse, who would drive around the houses and villages selling his goods. Later on, his mare gave birth, but he paid no attention to the tender little newborn foal. By then, as a small child, I had seen the lack of logic characterizing this phenomenon. It's the sight of an exhausted infant, trailing around them, tied to the carriage.

As could be expected, the infant couldn't keep up and died from being overly fatigued. The mother, who was active in her nature (did she even have an alternative?), became depressed. I saw in her body language the mourning, the lack of energy and will to live. Again, as expected, the owner whipped it aggressively in order to retain its livelihood. The horse died of heartbreak two weeks later.

During that same period of childhood, I witnessed the sale of a small foal. It was ripped from its mother and loaded brutally onto a crowded truck filled with adult horses. They pushed the foal by force into the traumatic crowdedness, the adult horses kicking the poor foal as it dropped. They ran all over it. It tried to lift itself up, and continued to be beaten and kicked. The money for the foal was counted and paid. "Polite" hands were shaken, and the truck drove away, with the faint cries of the foal heard in the distance…unto the unknown.

A man raised sheep for slaughter. One of them became prey to a bunch of dogs, but remained alive, with a dismembered body, which was being eaten by flies. When he was asked: "Why don't you slaughter it, and relieve it of its misery which is, after all, your aim?" He said in response: "I consulted my 'mentor' and he said it was bad luck."

Another example: people from one of the villages, who shall remain unnamed, told of the death of their mutual horse. They had starved the horse and abused it to control it more easily. They rode it one after another with each one pushing him around as much as possible in turn, to show he could do better. They have used a makeshift saddle made of an iron chair that hurt the horse's back repeatedly. His legs couldn't carry such a load. The poor and neglected horse suffered many injuries to its legs, and its back was collapsing. Its legs were collapsing. In fact, the horse itself was collapsing. As a result, its mental state deteriorated and it developed, among others, decay and pains in its teeth. The owners then hung the horse up, and pulled out its teeth, using a hammer and a chisel (without anesthetics). The horse died from the trauma, while still hanging.

Those owners have parents, brothers, relatives, neighbors and no one ever said anything. Not even the passersby who would walk by their home. Everybody considered it to be completely normal behavior. Such conduct is still practiced by the "human creature" without any disturbance. It's dark conduct.

An example of blind awareness and its ramifications: I'm talking about pallets on which goods are packed and shipped all around the globe. These pallets are made out of wood and nails. Many people, who throw barbecue parties, burn them, and the nails remain exposed for years to come. Every minded individual should think about the ramifications of their actions and make use of their wasted brain. Nails hurt people's and animals' feet, and endanger their lives. This kind of injury causes excruciating pain.

If we don't treat it in time, slow and agonizing death will follow. Furthermore, they ride their horses through the polluted pile of nails made by them, in a completely blind and atrocious manner. It bears particular notice that they don't draw any correlation between the horse's injuries and the burnt lands polluted with nails. In addition, thousands of forest acres are cut down daily to produce these pallets. These actions lead to environmental damage, the extinction of animals, and barbecue victims (i.e., the animals). Again, not only fixing the hole in the ozone layer would save the world. Flash was wounded three times from a rusty nail that was stuck to its hoofs (during the time I didn't have him). Therefore, we must stop manufacturing nails and glass bottles (especially alcohol bottles).

Flying colored balloons up to the sky creates a spectacular view. However, one must think about what happens to the balloons when they are in the sky.

Are we always aware of our actions? A family walks in a zoo. In one of the cages, a leopard is pacing aimlessly right and left. Many unaware parents say to their children: "Look at this crazy leopard that keeps walking backwards and forward." So the children learn and internalize the notion that tigers are mindless. But what if those parents were to use their brains and think of what goes on in the poor leopard's self" – being kept in a small cage all the time – and try to imagine its urge to break free, away from the surroundings of the dangerous humans. Then their children would learn to understand what is going on in their souls, and the distress of animals and humans alike. Only then will children learn from their parents how to logically and constructively think in the years to come.

Another example: people eat raw, tender, veal, without being aware of the calf's "self," which was poor and miserable after it was torn away from its mother at the time of birth. This agonized the mother. The calf was tortured when it was disconnected from her by force and kept in a small cage, in tough conditions, to soften its flesh and make it "more tasty and tender enough," to suit the refined palates of the eaters. The more the veal is kept in tougher conditions of captivity, the softer it gets (due to the stress it suffers and its weakness). Moreover, to weaken it more, they deprive the already miserable calf of eating and drinking for a few days before its slaughter. That is the aim, as a matter of fact. As far as the person is concerned, veal is nothing more than moving and living flesh.

If only those people were aware of the deep sensations of that calf, struggling for life, just like us, also wanting to run wild, play, discover interests, feed on tasty breast milk from its beloved mother and sleep next to her with a sense of assurance (because his happiness and self-confidence only lie with its mother). If a child grows up with these kinds of messages, he will learn how to use reason as an adult. And who knows? Maybe the next generation could save the world, crying "help," from the dangerous individual.

A clarification: milk belongs to mothers and their offspring. This is the milk that has been stolen from them by humans in their insensitive and atrocious selfishness using distorted methods. It is possible to stop the cumbersome culture of the dairy industry in exchange for plant-based milk, for example.

In case you don't know this, each day thousands of acres of land, home to animals for millions of years, are being "occupied." In addition, forests are destroyed for human purposes like: agriculture, industry, construction, and more land for herding sheep and cattle. Animals keep getting pushed aside, and more and more forests are cut down. Add to that hunting, pollution, and the drying of rivers, streams, and lakes.

If human beings were aware of our actions, then we might have a possibility of saving the planet.

"The child wants a pet." As far as man is concerned, an animal is bought as an object for his amusement. I was once told of a horse that was bought supposedly to use as a pretty, yet neglected, object to be kept in the back yard. And it indeed was neglected, lonely, and miserable, as many horses are all over the globe. A welding workshop was standing in the midst of that very yard, and the numerous sparks of light emanating from it caused the poor creature to lose its eyesight. This is a slow type of blindness, a burning one that is followed by a lot of pain. I should also mention that this story was told to me in an offhanded and mocking manner.

Yes, until writing this book, mankind (not all mankind) seemed to me a highly negative species, (to say the least) blind to its own actions and it included me. The common claim is that not everybody behaves this way. But if animals had an advocate to speak on their behalf, he would then say that many of his clients have seen that welding workshop, but no one has ever made a reasonable comment, and nothing was done to prevent it or to punish the stupid owner of the horse.

Almost every person in the world is allowed to purchase animals and take them for granted. We couldn't tell, up

until now, what horrible monsters we are to them...it's horrifying.

That is the limited and horrifying "Dark Maze" of humans. There are various foolish "theories," which are heard from these low-lives, regarding "animals." These "theories" add insult to injury. There is a deep and unseen gap between what really happens with an individual's attitude to animals, and the fact that that same individual cares and indulges himself, and supplies his own instinctive needs to his benefit (like drinking cold, clear water when it's hot, covering himself with an appropriately thick blanket when it's cold, walking out in the pleasant sun, taking shelter from the hot sun, endlessly varying his foods and more). Besides that, man was characterized by sadism, releasing his darkest frustrations upon those who are weaker (and whom he mocks), mainly animals, children, and women.

I hereby call again – we should immediately stop "taking" animals for granted. Animals aren't biological toys. Parents who get their child a chick or a horse as a toy to play with aren't ruthless. But let's put it in more accurate terms. These are inferior parents, dumb and mind-numbed. These human creatures have nothing to worry about because they are approved by a majority of our "justice seeking" legislators.

There are so many senior citizens who are left with no occupation, and plenty of unemployed and bored individuals. There are so many people who can make helping to save the planet a part of their occupation, if only we had the awareness to do so. It's possible to start a subsidized coalition and begin campaigning all over the world, in order to make a difference: to educate, to change, and to mend. Then there will be global harmony, world peace, and a more normal and stable life for everybody. Then we will live on a planet that is everyone's home, with less threat and more self-confidence. The possibility is out there and the sky is the limit.

SWIMMING TO THE UNKNOWN

Here's a moral.

(Flagellum: the male sperm cell)

> *Hey, Flagellum...!*
> *Yes..! You there, with the tail!*

Where are you swimming to so enthusiastically?

Toward the ovum?

Take my advice: perhaps you should think it over.

You don't know what you have coming your way. You are on your way to become a human being.

It's not easy being a human being.

You don't know who your parents will be.

There are different kinds of parents.

There are good parents, bad parents, and abusive ones as well.

There are parents who will dictate their beliefs to you, whether you like it or not.

You are nothing but a tiny creature.

You are dominated by older individuals.

They are not good or bad.

They are what they are.

That's because their brains don't dictate their ways and various methods.

That's why they are in big trouble with their big brains.

In addition, humanity is in its final stage of ruining your future home.

I don't mean to fill you with despair.

I just express reality as it is.

On second thought, I have written this guide book in

order to use the brain correctly. I hope this will help to save the planet (your future home).

If you're still interested in taking a shot, carry on and swim fast, because you have a lot of competition. Good luck!

TOURETTE SYNDROME

Tourette syndrome manifests itself by uncontrollable reflex-driven reactions of the body, like uncontrollable blinking or other involuntary physical reactions. Tourette is not a cerebral malfunction since it disappears during sleep. Therefore, the **reason** for the disturbance disappears in slumber. This is because in sleep we bid farewell to the daily reality that contains hardships and fatigue (see "**Sleep**" episode).

Conclusion

Tourette is a frantic disturbance that occurs among its owners, which is the "self." The various barriers standing in the way of expressing passions are manifested in a sense of "short circuit."

A breastfed baby is immersed entirely in his actions and

doesn't have spare time at that moment. Likewise, when a child plays a very interesting game, he leads a better sense of conduction. But in between, he stores dormant energies that don't manifest themselves due to hidden frustration somewhere in the subconscious like: desires versus prohibitions; a strong desire to beat what has yet to be manifested (thinking he's no good); a desire to be relieved of the sense of inferiority; and being polite with held back energies and not unloading the burden on others. There are many causes that spark uncontrollable behavior, which shakes the balance of the human condition. Sometimes a child wants too many things, but doesn't know exactly what. It's a frantic and obsessive desire, which yearns for conduction to fill up life and break away from restraints as much as possible.

These "short circuit" manifest in the body by the reflexes. The individual sometimes embraces these reflexes as body language to get attention (without being aware that is the case). It's a physical expression of its unconscious yearning to "purchase" the eyes around it, a feeling that says: "I'm being seen. I'm being thought of (for better or for worse), and I'm talked about. I actually exist" ("the quicksand swamp").

Various occurrences of this sort disappear and are forgotten.

In spite of this, a "lock down" might occur, which would later lead to an addiction to this "lock down," which then, in turn, would lead to a habit as a result of fixation. The inflammatory causes in this occurrence, according to the determining "self," result from the surroundings. Like: "You are sick." "You need a specialist." "You must be treated and rely on these potions." Sometimes a person acts like the class clown. He tries to fall in front of everybody and gets hit in the knee. That way he receives more glares from his company and he is supposedly more taken into account. All the above mentioned causes are circumstantial (meaning: a combination of various occurences - not neccessarily traumatic).

The treatment of Tourette is strictly psychological. You should detect the patient's frantic and hidden passions, and lift his "self" upward: from "inferior and sickly," to owning life by empowering self-confidence, so that he can express his hidden desires. Create a situation that will enable the patient to enthusiastically discover his new "self," and see life through different and rose colored glasses, as opposed to being held back by a restricting situation. I should mention that as long as the patient is still addicted to these habits, and doesn't (subconsciously) intend to get rid of them, no treatment will cure him.

Extreme treatment would include complete detachment

from the dizzying situation and the fixation he is in; complete disconnection from family, society, and the treatments he has been receiving thus far. Ideally, the patient should arrive at some faraway land that doesn't remind him of the previous dizzy spell. This is in addition to an occupation that leads to conduction.

It should also be stated that awareness of the causes of these occurrences is enough (according to this book). This awareness disconnects patients from prior conventions. The patient no longer has any possibility of blaming his innocent brain or body. In this new situation, he will find his individual ways for change and overcoming. This important paragraph is relevant to a wide variety of mental and physical illnesses alike.

FUNGI

The causes of fungi in the body are the above-mentioned ones ("the quicksand swamp" and "lack of resistance"). In order to get rid of the fungi in the fingernails, one must bid farewell to the direct and realistic look of the nail. One should divert his thinking by using the mental image of a healthy nail. Of course one should sense it at all times with self-conviction and a hundred percent intent. One should even take it further and tell oneself that this specific nail is thicker, prettier, and stronger than the others.

The same principle applies to every fungus in the body. We should all learn to use our divertive imagination toward healthy skin.

Moreover, we should improve our quality of life, since the quality of the resistance system is determined by the state of the "self."

ACNE

Acne is a phenomenon that matches a sensation. **It is not the hormones' role to create sores**. It's a state where the "self" has needs, without any options to choose from, without any preparation, or when it can't stand the pace. Not every passion comes into immediate satisfaction. There are general and sexual desires that don't manifest themselves, and even confuse and frustrate. The brain doesn't release hormones for us to be sexual (see "**Sex**" episode). During certain stages of development, every living creature was born into a sexual life and, feels a growing urge of sexual sense of belonging and a need to sexually connect with another (or others). In later stages, there is a need for continuity and then it splits up to produce offspring. There is no certain entity or brain to lead the boy or girl to cope with this challenging and confusing time. This competitive and demanding society demands that they handle situations they don't have the tools to handle. It's a kind of stress that matches in sensations the acne that is caused as a side effect.

According to the principle of the "quicksand swamp," we must not be involved in social competitions that convey the message, "They are more and I am less."

EDUCATION AND STRENGTH

To this day a child is expected to study all the subjects that were chosen as a goal to be reached toward the final exams. Unfortunately, there's no education on how to handle and succeed in life. A child doesn't have guidance to prepare him for life outside of school, at the end of school, or simply how to behave and understand others. In general there's no guidance on how to survive without despair and energy loss. The challenges of handling courtship, unrequited love, social constraints and their outcomes, are every child's bread and butter.

Guiding in self-expression with self-confidence, setting the goal of school and a program on survival in life, ought to be an inseparable part of the curriculum. Most children think that school is a fact, and that they don't have any other alternative. They even think of school as punishment.

Moreover, I would expect the educational institutions to add psychology lessons early on in school. It's the A B C of life. The school's psychologist should read the pupils' body language, especially during break time, to identify what's going on with each and every child, especially since they don't all express themselves properly by speaking. Some come from violent and traumatic homes, and the child is threatened and forced not to tell what he's going through. Some children control weaker ones and humiliate them, and no one seems to notice.

We should ask ourselves this: is it so important for the purposes of survival to study about Napoleon? To memorize the amount of people he killed? What he conquered and where? At which exact dates? Maybe it's more important to picture or learn how much damage wars have caused. How many mothers grieve and how many animals experience horrors (mainly the horses that took part in the excruciating journeys, were injured by arrows and spears, and suffered trauma!)

If we only invest in the child's schooling, rather than designing him, we miss the whole point. It's like learning to sail a boat, but not teaching it how to sail at sea.

SEX

Why is there a male and a female?

At the beginning, microscopic organisms lived in an orgy, with a sense of belonging and were dependent upon one another. At these stages, these creatures were multiplying due to their striving for energy and life, along with a lack of growth systems like digestive and draining systems, organs and senses. Therefore the "self" manifested itself. The "self" is, in fact, the "no choice" yearning and urging for energy, life, continuity, connecting, and multiplying. Hence, the ultimate "*I am my own maker.*" That is the reason why life is like the "quicksand swamp." A living creature rubs against another, and instinctively holds onto life, when life isn't taken for granted.

Take a newly-born living creature, crying out for touch, belonging and support, even when it's sated. When a living creature is on its own, it will feel helpless and lacking a grasp on life, since life is a given, "no choice" fact. Therefore, living creatures feel the instinctive and tangible need to be connected to others. A living creature lacks the instinct to multiply itself. It does have continuity and dividing instincts, but its senses indicate that two are better than one, according to an instinctive sense of belonging.

That's the reason why living creatures don't fertilize themselves. Again, there was no maker that made living creatures, ready with reproductive organs, in order to reproduce. I remind you, sex isn't for the purpose of reproduction. All living creatures are born visually sexual, with an instinctive desire to connect to others, not just mentally but also by touch. At later stages, the need is of physical connection, say sex, as a connection between two individuals. Making love and kissing on the lips are also types of connection and sex. Since every person feels lonely without someone else, the ultimate expression of connection is sex at an early stage, and reproduction at a later stage. When the male feels the urge to continue his lifeline by producing an offspring, he passes the female some of the essence of his life.

The female mates with another, who succeeds in making the impression of his courtship upon her, and with a sense that two are better than one. As mentioned, the body doesn't function by its judgment, but by sensations. The evolutionary romance equals the result of sexual organs, as an expression of merging and connecting, of pleasure and belonging, and later on as reproductive organs, for the purpose of continuity and splitting.

Therefore the question of who came first, the chicken or the egg, is irrelevant. In any case, the essence of our lives is based on sex, jealousy, and sexual possessiveness as an expression of belonging and being connected.

In the animal kingdom, a castrated horse, that is cut off from its accumulating sexual desires in a biological way, still carries on being sexual in a different way, but without accumulating relief in a biological and extreme manner. Sexuality manifests through pleasurable frictions with the other horses, or rubbing its teeth together. The horse is still longing for others. Yet there is an interesting occurrence: its desire for a female becomes less relevant, and is passed on to any good male or female friend. It's sex without sexual intercourse.

If, theoretically, we deny the castrated horse any contact

with his companions or his mate, it might end up depressed. It would most likely stop eating and its efficiency will decrease. That's because it doesn't have any grasp on life. In this condition, the horse is alone and without any social connection. In conclusion: sexual relief is a visual and accumulated connection in a biological manner.

When we add up all these parameters, we come to the above-mentioned conclusions. As a principle, when the goal isn't reproduction, sex doesn't have to be between males and females, as a rule or convention. Animals boast about their sexuality and stress it. Man, unlike animals, covers himself with clothes, to hide his, allegedly, embarrassing sexuality. This is the source of sexual complexes.

Despite what is said, in the human culture, a principle of black versus white has evolved. These are strict laws regarding sex. Sex is a very possessive field. There are those who are a product of their culture and act without knowing why they are doing so. It is for that reason that we find cases in which an average individual preaches against homosexual relations out of various and subconscious factors, like sexual jealousy and possessiveness. But then, in a twist of fate, that same individual can find himself in such a situation as prison, where homosexual intercourse is common, and he may take part in the rape of another man. Moreover, his friends might join him in a gang rape. All this could happen and has happened to people, even at times when this sort of sexual activity has been a taboo, even as far as that very individual was concerned. Unlike animals, this hypocritical, poor, confused and primitive conduct exists only among man. May I remind you that sex cannot be performed without an urge, nor can it manifest itself without the urge that is sexual attraction. This field involves a lot of hypocrisy (only among man). In conclusion, sexual preference is a matter of pure taste, with no conscious, will-driven, birth right choice. It is similar to a living creature that is born with a sense of taste, but is still completely individual. That is the infinite "self."

Some prefer cherries over strawberries while others prefer the opposite. It's subconscious, individual taste and it is not genetic. There is no guiding hand or any type of entity that guides the individual sexual attraction. An extreme and violent homophobe has homosexual attraction in denial, who is forced to declare to those around him that he is not a homosexual." This growing phenomenon has become a spinning trap man has unconsciously invented for himself. There's only an instinctive sexual attraction. Therefore, sexual preferences are operated by an infinite variety of means. That is the difference between plants and living creatures, which as far as we are concerned also includes humans, who are creative and complex creatures.

Each one views the issue of sex through different eyes. Therefore, **sex is an infinite and individual sensation of the "self" according to its taste**.

A homophobic person is someone who is frustrated with jealousy over the oppression and limitation of his own sexual freedom and its diversity. It's the effect of his surroundings that have pushed him to the relentless social convention. It's highly likely that an extreme and violent homophobe has homosexual attraction in denial. This individual is forced to declare time and again to those around him that he is not a homosexual. This growing phenomenon has become a spinning trap man has unconsciously invented for himself.

The issue of sex is driven by instinctive and possessive sensations. These sensations are manifested through subconscious jealousy; and therefore, the strict rules (that were imposed by men, who have for centuries been the dominant sex). It's that latent sense of jealousy that says: "I follow these rules and prohibitions obsessively, and I pay for it by not fulfilling my forbidden desires while others enjoy sex more than me by denying these prohibitions."

Man was the one limiting himself by these rules. Those are individual sensations given by individual interpretations.

A demonstrative example: No one cares whether this person prefers cherries over strawberries, because the sense of jealousy doesn't manifest here, nor does it create a sense of emotional burning.

In conclusion, each living creature is born with sexual instincts. That is the basis of life. There is no particular entity to direct us on how to specifically use sex.

There are also social conventions that determine that we should produce as many children as possible, regardless of the results, claiming we do not call the shots. This convention brought with it some grim results like an over-populated Earth, plagues, wars, famine and extinction.

One child is an entire world, so we shouldn't treat him or her like they are merely a part of a herd.

Throughout history, certain sexual prohibitions were created that have become over time legitimate. Disturbances, confusion, and misguided approaches to sex all take place in the "self," in an individual manner. The biological barriers are there, and shouldn't be blamed for being the obstacles and restraints to the quality of sex. Since the body works by the "self" and not mechanically (sexual organs have their home owner, the "self" who operates them exclusively, in an infinite and subconscious manner).

The will doesn't call the shots during sex. The issue of sex is only operated by infinite sensations of its owner, alongside his interpretation of the changing situations taking place throughout life. For example, an individual confronting his sexual attraction. Perhaps the sexual attraction that this individual felt was too extreme, and unbecoming of the situation and his will. Or rather, a homosexual attraction, which is unaccepted by him and goes against his will, and according to his interpretation, that is driven by social convention.

Sexual arousal doesn't consult the wishes of the body's owner. The will, for itself, is separated from the individual sexual orientation. Every living creature is born by an instinct of yearning to belong, to connect, for the sake of continuity and multiplicity.

These sorts of occurrences happen by the "self" in a subconscious manner, to each living creature individually.

Don't blame hormones, since the hormonal level is set by the sensations of the body's owner.

The rules, strict laws, and prohibitions regarding sex are set by man himself rather than logic. They come from the depths of the jealousy-driven sensations, according to infinite and individual interpretations. Yes, some of them are very twisted rules that can make you sick.

In fact, all the rules throughout history (not necessarily regarding sex) were set by men. The passive women had to accept them, which was what created the fixation. Change will only occur if a reaction is being manifested in defiance by the no longer passive women.

Though man is the creature with the largest and most complicated brain, humanity didn't receive any instructions on how to operate it. And this has gotten us into quite a jam. Without instructions on how to use this "super computer," the "self" is left unguided and might get lost. I hope this book will become an ultimate guide on how to use the brain correctly, making the most of it, in order to improve the quality of life and save the future of the planet.

LOVE

There are different kinds of love; but all are expressions of emotional connection with someone else. There's love between a parent and a child, and vice versa, love between siblings, friends, between men and women, and even between mankind and animals.

Love between partners is expressed through the "self" that is crying out for intimate and emotional connection, and for a sense of exclusively belonging: of two feeling as one an inseparable connection, by an individual impression.

It is an obsessive joining of arms, which ties the two beings into one.

The body doesn't fall in love with another body; the brain doesn't fall in love with another brain. A body doesn't fall in love with a brain; a brain doesn't fall in love with a body. It's the "self" coming together with another "self." For a while, throughout the courtship, they might merge, and connect more and more visually. The body will behave according to its sensations, for better or worse, without any "emotional guide" (the brain doesn't project feelings since it is actually mindless). It is impossible to fall in love with an object (for instance, a pretty doll), as it does not have a "self" to be charmed by or to harmonize with.

Love has nothing to do with willpower, need, ability, biology, obsession, rewarding, bribery, flattery, judgment, begging, or showing affection toward another. Love is made when one side's courtship assembles and assimilates with the courted side, and the courted side is impressed and visually connected to the courtship.

These terms aren't written in our brains and are not guided by any source. It's a gentle and delicate art. Each suitor should know when the person being courted is retreating. We should give him or her space, and not "strangle" the relationship by routine, over-flattering and self-degrading. The suitor must always maintain the appropriate intensity of courtship, and always leave the other wanting more. Not every person is suited to the same type of courtship.

Sex between lovers is the fulfillment of the sensation of a visual connection, through the connection made, and the fulfillment of the desire in a physical and tangible way. That is one of the highlights in the life of a person in love. However, this is not suitable for everyone. Each person is a different infinite individual. Therefore, unrequited love could be the fate of the best suitor in the world. In this situation, there's only one choice, which is: to leave. The more a person in love suffers from unrequited love and carries on courting the other, the farther he will sink into humiliation, emotional, and physical shattering, lack of energy and actual collapse, and even suicidal thoughts.

There are rules to courtship that we should all be aware of. For example: courting a girl who's distraught due to a lost love. It's a dangerous step that doesn't stand a chance, because as far as she is concerned, her "self" is still connected to the "self" of her loved one. There is also a possibility that she will develop a drive of revenge and contempt toward the offending sex. This will be a frustrating courtship and offensive to the ego. It's a courtship that is doomed to failure and crashing, according to the "rule of the scales."

When the "self" is mutually in love, the body produces positive substances depending on its sensation.

When the feelings of both parties' "self" aren't mutual, and the courting individual feels consequentially humiliated, his body produces negative substances according to his sensation.

It is the "quicksand swamp."

The body doesn't produce feelings by chemical reactions. The spreading of this wrong information should stop.

THE CONNECTION BETWEEN JOINTS IN THE HUMAN BODY AND SHARKS

Everybody knows that the joints are made of cartilage. Some people have therefore (blindly) assumed that if sharks' cartilage was somehow infused into the human body that would solve joint maladies. Some people eat soup made of sharks' fins. And for those of you who don't know, many sharks are brutally murdered in the process. Their fins are cut off, and they are tossed back into the ocean while still alive, and are left to solely and agonizingly die.

I hereby unequivocally declare that sharks' cartilage does not help to heal human joints! (See "Arthritis" episode).

Another example of this sort of damaging mistake is the claim that Omega 3 improves the brain's efficiency and that we should therefore eat fish.
I hereby declare that Omega 3 does not help or improve

the quality of the brain. In fact, Omega 3 supplements cause damage and poison the body (See "Placebo" episode).

I turn to you again: save the fish from the intensive fishing, because our world is disappearing right in front of us.

Did you know that the seas and the lakes, that look so beautiful and spectacular from up above, are actually humanity's and fishermen's garbage can? On top of that, many fishermen throw away their worn-out equipment into the sea. Then sharks, dolphins, whales and other marine animals are caught in the nets, and drown slowly and agonizingly. At times, they may be caught in the rusty hooks, and die slowly and in great pain. It's a never-ending wheel of torture, since other marine mammals bite the dying fish with the hook inside it, and so on.

An awful disaster occurs when sewerage, toxins, and other chemical substances are poured into the seas and other water sources.

As for fish pools, it's natural for migrating birds to feed off these pools since they don't understand they have entered a land owned by mankind. In order to protect the pools by using nets, the birds are shot with hunting rifles.

Does a fisherman love fish? If so, then why the hell does he eradicate them with such easy pleasure?

Is mankind such an awful evolutionary mistake?

As long as this behavior continuous uninterrupted, we are all accomplices to a crime, since we are standing by and making footnotes, in the hopes that there is an entity that is in charge of what's going on. The internal sacred order calls upon the adolescent youth: do some deeds-! Because nowadays, mankind is satisfied with mere talking and idle chitchat. Don't be passive! Stop consuming the so-called "healthy substances" and idly crossing your fingers over your satiated bellies! Be adamant with your goals and keep your only home existent in the universe. Hold your heads up high and hit the road. You are our only hope.

FOOD

The food that we eat not only contains all the essential nutrition and energy, but also supplements and toxins. These are drained by the *"Ultimate I am my own maker"* according to its senses of conduction and lack thereof, for better or worse. The repetitive claims that some foods are "healthy," and therefore, we must consume them to sustain and improve health, are wrong and misleading. As a result: some people eat not in order to satiate their appetites, but only for their health. In addition, the amount of food consumed is a lot more than what a person needs, under the pretense that it's "good for our health."

Conclusion

Don't try and be smart with the laws of nature. Eat by your individual senses of taste, hunger, and satiation. That is, according to personal sensations. We'd better part with the hobby of eating and learn to eat a little at a time, and once we feel satiated we will fly like birds.

Overeating doesn't benefit your health, wisdom, or sexuality (to say the least).

Here's an example from a scientific television show: a girl whose size is rather small for her age. The girl's parents try to force her to consume more proteins and vitamins, in order to induce her growth. But when the satiated girl wouldn't eat, her parents would punish her.

In the biological field, mankind is progressing backward. **And again I declare that vitamins and obsessive overeating don't benefit the acceleration of the body's growth.**

This sort of behavior exists among modern societies that enjoy an abundance of food.

We hear daily phrases that make no sense such as: "Dairy, desserts, and snacks are not food." And "You must sit down, and eat normal and nutritious meals."

Well, the dairy products are made of milk and that feeds all mammals. Snacks are made of foods like cereals and legumes. Therefore, the conclusion is clear. It should be noted that when a baby stops being breastfed, his body no longer needs milk or dairy products. Therefore, milk is absolutely redundant.

The more you push a child to eat, the more his obsession with repulsive food will grow and develop into a state of nausea. Henceforth the decline ensues: like digestive issues, weight gain and even anorexia. Modern man is in a dilemma, resulting from a twisted interpretation given by laboratory studies that are embraced with no questions asked. Like "the functional development of the brain"; "growth and development"; "the dangers of lacking foods that cause diseases"; "if you don't eat calcium-enriched food, it might damage your bones by the time you are ninety"; "calories count" and so on.

The body has its own pace of growth, regardless of any nutrition it receives. Only in an extreme case of hunger will the growth be slowed down. That being said, half of the world is starving, and half the world is fat due societies' ever-changing mood swings and fixations.

A car consumes fuel according to time and speed. Human and animal bodies don't work by the same principle. Not according to a predetermined time, or by the effort, or number of steps. The role of the personal senses is to determine what to eat and how much. Overeating is a result of various factors like over-compensating a non-compensating life, a twisted cultural perception, laziness, and an inferior personality.

A side note: race horses, for instance, can be satisfied with light hay as nutrition (especially when it's hot) because these horses have about twenty-three hours a day to eat or get tired and bored by the frustrating routine (inside the stable), unlike horses that labor throughout the day. Unlike racing cars, a shot of energy is exhausting for the horse, regardless of the type of food. Adding a little tasty mixture improves the sensation (when the quantity is more than a little, the reverse effect is created, despite what we would have expected to happen). A message to nutritionists everywhere: it's about time to stop with the magic solutions of condensing energy into living creatures. These are miscalculations that occur constantly, even now as I am writing this book. These ignorant calculations miss the point (See **"Urge and Lack of Urge"** episode).

The television remote, for example, wasn't made due to lack of biological energy or in order to save food and energy. It is the "self" lacking desire and energy to operate its body, regardless of any type of food whatsoever.

WHY...OUT OF SPITE?

Why does man find himself in the situation of mental and physical collapse? Why does everything seem to be going wrong at times? One explanation is this: it's the "self" determining to move in a better or worse direction, in addition to a chain of reactions. The brain has no intentions, neither good nor malicious. For example: a person hurts his toe and is afraid he might be hurt again when someone accidently steps on his foot, hurting his toe. All that the "self" fears or hopes for can come true by self-guidance, for better or worse. If a passerby does end up stepping on his injured toe, then the "self" has actually fulfilled what he had feared. This realization is traumatic for him, partly because of the frustration of pain, but mainly because of the over frustration of "Why, out of spite?" (meaning, frustration leads to another by subconscious self-guidance to rock bottom).

This insight is relevant and applies throughout the course of life, and not just a specific event.

I should add that insignificant and unnoticeable situations pass us by without us paying any special attention to them. But highlighted scenarios, whether good or traumatic, burn our consciousness more vigorously. Therefore, there are leaps to the top in the best case scenarios or a decline to the bottom in the worst cases. When a man believes in a guiding phenomenon caused by someone out of spite, it can lead to a chain of reactions, according to his interpretation, whether declining or rising. These endless dramas take place by the infinite "self." It is the "quicksand swamp." When a man feels like he's drowning, he sees the world as a dark place. He then either pushes and shoves his way up or loses himself, his senses, his hope, and life's energy. In this situation, the supposed "quicksand swamp" sucks him into the deep water. See **"My Story"**. It's a moral that says: "Use your brain, especially at times of crises, because there is no one to pull you out of there and think for you. The brain doesn't think for you; it is the "self" that does so."

ANOREXIA

If animals or humans were biological machines, we could have injected food through our veins and saved the body from starvation. But the body has an owner. It is the "self" that runs the body and no other factor. Therefore, if we inject the body with food while its owner rejects it, the injected food will become irrelevant.

The media's daily talk of food leads to the public misconception that body weight is a direct result of the amount of food ingested and its ingredients. An anorectic girl, as could be expected, operates her body according to this belief. The body is run by an individual interpretation made by the "self," so we must remember that there are skinny "eaters" and on the other hand, there are fat, hungry people. There are skinny people who have gained weight without increasing the amount of the food they ate. There are also fat people who have lost weight without decreasing the amount of food they ate. The internal sacred order calls again to stop this "ill" chitchat because the public conducts itself based on it, and that causes a lot of frustration and tragedy. The anorexic girl should immediately step out of this deadly dance of demons, look at it from afar, gain perspective, and feel nothing but utter contempt for it. We should live in complete disconnection from this moronic and frustrating cycle. Anorexia doesn't exist within tribes whose nutrition is minimal and unavailable.

There are also cases of stupidity. Yes, man is the only creature who has a choice to act foolishly (not being aware he is doing so). He bows his head to society; learns to imitate it by observing it and following its rules; and acts like an inferior robot. Phenomena like mindless discussions (food, vitamins, calories etc)., twisted behavior, sick body language, sickly conduct and unhealthy thinking all exist only among humans. All the above reasons are coincidences that oppose survival.

Take a girl who weighs twenty-five kilograms and is still convinced she is fat. She is definitely the slave of the twisted information that she received. This comment is relevant for a wide variety of both physical and mental occurrences.

Each occurrence or obsession has a beginning. An obsessive phenomenon can lead one to a state of dizziness, and either a slow or a rapid collapse. Because the body doesn't operate by free will, the more the anorexic girl is pushed to eat, the worse the obsession within her becomes, and her rejection of the food grows. That is because will is irrelevant. This is an individual interpretation. Hence, a good psychologist alone should examine the subconscious causes of this phenomenon (or any other obsession), each case on its own, and change the interpretation with logical steps. Not by trying to convince, plead, beg, defy, accuse, show mercy, or ask for it, but only by solving the psychological puzzle completely. When a small part of the puzzle is missing, all methods are completely useless to him. It is the highly complex and infinite "self."

A concluding sentence: dear mother, stop shoving food down your child's throat. It is a biological toxin in the short term and a mental toxin (the "self") in the long term.

PRINCIPLES OF OBSESSION

Every phenomenon has a beginning. Meaning, an obsession starts with an individual interpretation, a reaction to the unknown that leads to a collapse. A reaction can occur in a blink of an eye or after some time. For instance: a man insults and humiliates his partner. She reacts by instinctive emotional "drowning." Then she tries to fight by "pushing and shoving" her way out of this situation, to try and perform a sort of "counter offense." The man, as is widely known, is more sensitive to the injuring of his pride, and so he immediately reacts with violence, according to the "scales rule." His partner then finds various ways to get even with him. And thus continues the vicious circle to the point of collapse.

Another example: a lot of deadly road accidents occur when the driver senses danger and imagines the worst of all in the blink of an eye. All that the "self" thinks and imagines will be realized by it; meaning the "self" driver is "controlling" the car by the interpretation it makes in that split second (See **"Concept of Time"** episode). We should be aware of these laws of nature in order to prevent tragedies from happening.

Two boys meet in a street fight. One makes an offensive comment. The offended person draws a knife to reclaim his inferior pride. The offender also instinctively draws a knife, to try and save his honor. Actually, each can run away and save his life, but a reaction is created of forces binding them to this momentary tragic situation that is carried out naturally by the "quicksand swamp" and "scale rule" principles.

The individual "self" is caught in an obsession when its interpretation and annoying sensations are not resolved. Obsessive cleanliness: this phenomenon is an individual interpretation and not a cerebral malfunction. This (supposedly) "sick" person is convinced that there are germs around causing illnesses. His sense of rejection forces him to get rid of their presence by an uncompromising form of maintaining cleanliness. It's legitimate as far as he is concerned. To change that twisted interpretation, one requires a new and reasonable explanation for the phenomena.

Before the microscope was invented, obsessive cleanliness didn't exist because men knew nothing about the existence of germs. In fact, most animals have survived and still do, to this very day, not knowing of the presence of germs and they are better for it (the appearance of germs and the issue of resisting them is detailed throughout this entire book). When we examine the principles of this occurrence, we actually discover that it is not an obsession at all, as we have thought so far. Because diminishing and even preventing ailments (according to the person's individual interpretation) is legitimate.

MEAT

One of the main goals of this book is to stop men from eating dead animals, and stop the enormous massacre of animals and the scavenging. Most importantly, men should be aware of their actions and their ramifications. We must save the planet from extinction.

A worrying discovery about meat

In order to produce one kilogram of meat, farmers feed the animal with human food such as cereals and legumes, from the day it is born until it is slaughtered. The poor animal is stuffed with food that weighs at least twenty times more than the animal itself. And let's not forget all the other hardships that captured animals face such as death, abortions, diseases, plagues, a mother that was malnourished while pregnant, etc.

Conclusion

The investment in meat is enormous. It takes away men's food and agricultural land. Therefore, if we use this food for feeding humans, we make twenty times as much, and as a result would also:

1. Stop the enormous massacre of animals and become aware of the injustice that has been blindly executed.

2. Stop raising domesticated animals to be led to slaughter.

3. Decrease the phenomena of over-grazing. As a result, there would also be a decrease in the destruction of forests (which are being cut down for herding domesticated animals).

4. Increasing the amounts of food for people all over the world, which would also partially solve the starvation problem.

5. Begin Ethanol production out of corn rather than fossil fuel. This would be a better use of corn than feeding it to animals.

6. Decrease the emission of greenhouses gasses that are causing climate change, the expansion of the deserts, famine, starvation, and wars over food sources.

7. Significantly decrease the traumatic transportation of animals by trucks, ships, trains, and journeys on foot.

8. Decrease the spreading of diseases caused by the crowding of domesticated animals, like SARS, Nile fever, the avian flu, and tuberculosis.

9. Significantly diminish the number of flies and mosquitoes passing on diseases, consequentially drying up the seas, streams, and lakes (caused by the excrement of domesticated animals).

10. Stop the massive amounts of injections, like antibiotics and other substances, administered to animals during sea voyages to prevent or treat nausea and diarrhea (they suffer enough horrors on the way that are only intensified by the disrespectful treatment they receive from their caretakers).

11. We will save (rid ourselves of) the stench of the carcasses and its implications. These implications include damaging morale, romance, the quality of sex life, etc.

There are some other advantages, but there's not enough room to count them all. The purpose is to encourage man to start being aware of his own actions.

A lot of predators are shot to death by farmers – wolves, lions, leopards, and cheetahs. The farmers claim that they are endangering their herds and cattle. This is one result of the occupation of man and the extinction of wild animals. The Tasmanian pocket wolf, for instance, has already been made extinct by farmers.

Animals that are taken to slaughter should not be treated in such a demeaning way, being transported by trucks, then trains, then ships, then trucks again and so on. This is highly traumatic for them, so at least up to the point of slaughter, they should be treated in a more humane way.

As for the leather industry: man sees animals as inferior beings that exist for his sole purposes, since he is the noblest creature of all. A lot of idiotic expressions come from the perception of a man like: "They are made for it"; "It's their calling"; "It doesn't bother them"; and "It makes them stronger." Farmers maintaining cows in compounds keep the cows' heads secured between iron bars so that they can be forcefully fed, thus increasing the quality and quantity of their milk. In this situation, the cows can't banish pestering insects from their bodies. Their skin then becomes all bitten and swollen, and eventually worms are drawn to their festering wounds. A source in the leather industry was quoted as saying: "The leather industry loses billions each year because of the poor quality and bites in the skins of the cows." The cows' agony is irrelevant to them.

Products can be made without leather, such as light, flexible, more comfortable and healthier shoes, so what is the purpose of this fixation on leather? Can someone answer this?

Foxes are another example of animals kept in tiny cages

under the open skies in harsh cold weather conditions, in order to speed them up into developing a wintery fur, and make it long and soft enough for the fur industry. And these furs are also used to make, among other things, religious accessories.

And that's not all: the poor foxes are skinned alive by human scum. Heaven forbid we should waste a bullet on "these objects." We waste no time killing them because the buyers must be supplied with perfect fur.

Society should frown upon those perverted people who would, despite this classified information, continue to buy fur products.

I would like to take this opportunity to call biological scientists to come up with a humane method of putting animals to sleep: a method that would save the pre-slaughtered animals' traumas, and spread it throughout the world.

By the way, the act of slaughter was derived from an ancient belief, which prohibits drinking blood. This exhausting and excruciating process was set to drain the blood off the animal while it was still alive. Animals are the victims of this belief. Those of you who are convinced that animals don't understand or that the slaughter isn't traumatic for them, or that they don't feel, or that they lack our will to live, or that the post slaughter spasms are mere reflexes – all you wisecracks would be better off trying to open your minds and see the world from another perspective for a change.

The irony of it all is that man accuses the wild animals of spreading illnesses among human beings, and because of that he restrains the wild animals and even kills the wild flocks, which were actually infected by domesticated animals. The root of the domesticated animals' illnesses is their great crowdedness and living conditions that didn't exist throughout evolution. Stress, crowdedness, and routine bring new diseases along.

Setting boundaries for wild animals only adds to their stress levels and causes other disturbances such as: disturbing the migration of birds which is necessary for the collection of food and water, which then leaves the birds hungry and thirsty. Moreover, without migration, both birds and cattle have no choice but to mate within their own groups, which then causes genetic deterioration from one generation to another. The genetic lack of diversity begins to snowball, leading to the birth of more and more sickly animals that are less immune to diseases as the generations progress.

The wild animals are infected by the domesticated ones,

and spread their illnesses from one to another or by migrating birds. That's how mutations of new illnesses, that haven't existed in previous stages of evolution, are created.

That's why I shall make this unequivocal claim: human beings are accountable for new diseases and plagues, since we are the ones who twisted the laws of nature.

As a result, a chain reaction occurs that gradually disappears from our sight. Over-herding causes the edible grass to slowly disappear. Here the "rule of scales" manifests itself in weeds, which like inedible thorns, take their place and so expanded deserts of thorns are created, along with famine and extinction. It's a boomerang that affects humans as well as domesticated and wild animals. A chilling example to demonstrate the ramifications: the amazing cheetahs accelerate the speed of their running by manifolds and reach an inconceivable speed as an existential strategy. At this speed, they run into thorns that injure their eyes. They are blinded by them. As you know, a blind cheetah cannot survive, meaning they are becoming extinct. Their extinction leads to more animals becoming helpless prey. But they still don't have any edible food other than the prickly thorns, so the prey is thus extinct as well...and so on. Add to that the global warming mentioned above, and the result is quite obvious.

Were there some sort of guiding and creating entity, would it have given us a free hand in continuing our looting of its creation?

Despite all that, man still eats animal carcasses, and so I give you a recommendation for a blessing to say before meal time: "I'm hurting and deeply sorry for raising you in such horrible and demeaning conditions, simply to satisfy my desire for carcasses. Forgive me that your pathetic life was bluntly taken away from you – for my benefit." No need to say, "Bon appetite." There's really NO need!

In conclusion

We human beings will feel better about ourselves once we stop with this mass slaughter of innocent animals, turning them into carcasses for feeding human beings. It will be a wise progression toward rescue in general.

The human creature must at last be human.

I ask science in general and biology specifically to stop "poisoning" the world and men with meat supplements, vitamin B12, iron, protein, Omega 3, various interchangeable chemical substances and so on, because it contributes to extinction in an indirect and subconscious way.

And what about the irreversible damages caused by extensive fishing?

Man calls it fishing, but it's an enormously massive slaughter and the extinction is already apparent.

I repeat my unequivocal statement: eating carcasses is completely redundant. The human body doesn't need it. There are whole powerful tribal societies refraining from eating animal flesh. These people live a stronger and healthier life than the "modern man" does, and they even live longer.

Furthermore, the predator's body doesn't need carcasses specifically. It's that their digestive system (which has been designed throughout evolution by the "no choice" element) isn't built to process grass. The body needs energy, but it doesn't matter what source it comes from. Animals have evolved from eating vegetation. The prey is fed by vegetation and the predator is fed by the prey. The basis of energy, from which we develop, is in fact vegetative. Milk comes from it too.

I hereby call food manufacturers to urgently invent a formula for vegetarian food that tastes like meat; and it would be even better if it could taste like tender veal and fish as well. And on a mocking tone: to quiet the nutritionists, please add a lot of vitamins like B12 and iron in a placebo hoax.

By the way, there are vegans/vegetarians who suffer from shortages, but the reason doesn't originate directly from the food they eat (see the **"Anorexia"** episode). The reason is completely indirect. It's the agonizing and anxious individual who averts the horrifying ways of carcass eaters (and that is legitimate and human). Bear in mind that there are vegans/vegetarians who don't suffer from shortages. Moreover, there are carcass eaters who suffer from shortages.

A message to the wealthy, western countries, careless of what's going on in the world: if you don't start immediately by changing and installing a new regime based on the recommendations of this book, there will be a deadly chain of reactions, an uncontrollable collapse. For instance, immigration from undeveloped countries to developed countries might exceed all resources and get out of hand, because as far as the immigrant is concerned, it's a choice between life and death. There will be endless civil wars causing turmoil, massive looting, violence and crime.

A recommendation for biological scientists

I should mention that science in general and biology in particular saves lives. Despite the unanimous agreement among specialists from the field of natural sciences, they still haven't formed a statement about the existence of a deformed missing link. Science practitioners must declare that they have reached a dead end and they shouldn't draw any "magic solutions."

Moreover, it's time to stop conducting unprofessional studies that are now reduced to mere conclusions. New studies keep on contradicting existing and previous ones over and over, without any logical scientific basis. It should be noted that the public feeds off the science of biology completely, conducting their lives according to new biological scientific studies that contradict their predecessors. It's confusing and frustrating. Instead of driving the public crazy, biologists should save on these wasted and frustrating energies, and invest them in saving the world.

It's time to stop preying on the bodies and innocent minds of wretched lab animals, and alternatively, start using our brains to a greater extent.

These horrifying and stupid experiments continue. Man is

getting lower and lower. For example, baby monkeys are brutally and forcefully taken away from their mothers. At the beginning of the experiment, scientists put the baby monkeys through a lot of misery. They frighten them in order to test results. For instance, do they prefer a cage with food over a cage with a doll in a form of a mother monkey? (Obviously, they would prefer the mother monkey). Then, the scientists put the baby monkeys in complete isolation to test results. Then, they put each baby monkey in a deep dark pit, all alone, for a year, to test biological phenomena like Alzheimer's or other brain illnesses.

The babies who survive the inferno go through more experiments [see **"Religions and Beliefs"** episode + Acceptance (subsection in the **"Witchcrafts"** episode)].

Dr. Josef Mengele, the notorious human filth is living, breathing, kicking and duplicating himself over and over again.

The invention of the microscope caused men to conduct their lives according to the laboratory results and give up on the logical mind.

Vitamins

Vitamins, in a small dosage, aren't harmful because they are absorbed into the body. And if it is true that vitamins are more suitable for a hungry body, why don't governments then distribute them in starving countries?

In order to prevent deforestation, I recommend a formula of material, which will replace the use of wood: sand. Yes. Sand + flexible adhesive in different colors.

TRAPS

Strangulation traps are made with barbed wire and loops in order to capture animals. How would you like to watch a documentary about a lion or an elephant being captured? It's nothing short of an agonizing death.

If the elephant, for example, succeeds in escaping the trap, the loop will get caught on its trunk or leg. And the animal will remain trapped and eventually, after long and excruciating agony, suffocate while other animals in the herd will only stand by sympathizing with its helpless sorrow, unable to do anything else. Sometimes, the situation is made worse when the captured animal has an incredible urge to return to beloved and hungry offspring, who are waiting for their mother. It's a sensation of great horror, up against cruel fixation, that says: "They're going to eat me while I'm still alive," missing an instinctive reaction to defend itself by fleeing.

There are countries where certain animals are already extinct (our world is gradually disappearing).

Did you know that in the past, some of the antelope species migrated throughout the Middle East? Humanity has completely eradicated them, using sadistic methods. The hunters pushed them into a contraption resembling the neck of the bottle, and at the end of this contraption were wheels with blades on the edges. Each time the antelopes made a pathetic effort to exit the bottleneck, they were severely injured. Some of them died a short while afterwards, and some experienced a very slow and agonizing death while their bodies were dismembered.

Often at a barbecue, we would hear typical things, such as: "I heard that Thai people eat anything that moves."

This sentence is said when logic, imagination, or awareness concerning the brutal and agonizing methods of animal murder are lacking. The more helpless the hunted creature is, the more tender its meat. Moreover, most of Thailand's landscape was composed of vast forests that are now completely destroyed.

And from the animals' point of view, we can almost hear them say: "For us animals, men are an awful tragedy. We are aware of our agonizing and tiring defeat. But we wish our defeat would not be so slow, but rather occur in a blink of an eye.

"Our lives are hard even without you, humans. We barely survive as it is. Please, leave us alone and prevent our creation. You can do it. Even when we have a good and humane owner, most of our lives are forced and boring."

Dear Parents: Please stop instilling in your children distorted conventions such as: "It's their job." "That's what they're created for." "They are built for it." And "They are created for mankind."

The internal sacred order calls to all you who are traveling to India to experiment with various drugs that you have been told possess the power to help you step out of reality and twist your rational logic:

Burying your head in the sand is dangerous and irresponsible. You should keep your eyes open to reality and do deeds that benefit the rescue, rehabilitation, and design of the globe that is your home.

Specifically about India: haven't you noticed that India, like other countries in the world, is becoming a trap of garbage? It will collapse by its own people, who multiply uncontrollably. Doesn't this phenomenon cause your insides to turn over? So our top priority as far as third world countries are concerned is spreading the message that overpopulating the Earth is negative – babies shouldn't be brought into the world when we cannot support them. What have they done that they must be punished by a life of outrageous poverty and misery? Alternatively, people should focus their energy on raising one successful, happy child. It's a sacred internal order that is humane and is directed to any person in his right mind. Heaven is only as far as you choose it to be (it's not far up in the skies).

MARRIAGE AND FAMILY

Is it mandatory to get married because that's what our ancestors did?

To bring a child into the world, you don't have to get married. Most emotional relationships don't last very long to begin with. Love reaches its climax, and from that point on there is a slow or fast deterioration. This happens for various reasons: routine, boredom, a match lacking interpersonal communication and much more.

Both parents (who, may I remind you, are not married) can decide whether or not they want to live together – according to the guidance of their instinctive sensations (love, sexual attraction). That way they will remain young both in spirit and body.

A marriageless family should bring only one child to the world, who will allow for a fun and satisfying dynamic in the life of this family, unlike the burden and responsibility experienced in families with many children.

People who get married according to fixed and written norms start aging from the beginning of their marriage. Once a couple is married, they plan their future and end in a subconscious way. Meaning that the goal is: to bring children to the world, grow old, age and be cared for and supported by our children.

An only child is a privileged and more successful child! He's more valuable, more invested in, and gets more love, attention, education, and material things. He would feel less inferior without another brother to compete with and humiliate him.

We should make it obligatory for parents to take a parenting class even before they have brought a child into the world. There are no instructions provided by our brain to guide us on how to take care of and educate our children. This kind of class will save children (as well as parents) a lot of heartache.

I hope this will be read by leaders – people who have their heads on their shoulders. People suffering from fixations won't understand the essence of the book unless they are brave enough to part with them and open up to new ideas.

RELIGIONS AND FAITHS

There have always been leaders. Even among animals there is a dominant leader or a matriarch acting as the "wheel" of the herd. However, among humans this role has developed to an unpredictable and disproportionate size. As could be expected, no leader has the solutions to all the ailments of dominant people. Therefore, the road for various faiths and dependence upon imaginary entities is inevitable. Later on, these faiths turned into legitimate and official religions.

Faiths and religions were created as a result of the lonely

state of the human being. Being alone without someone in charge is sometimes an insufferable reality, reaching a mental rape in response to the helplessness of insufferable situations.

If there were an entity managing our lives and our world, then the world would be clear and factual, and there would be no need for religions, faith, and administrative entities. Or, alternatively, there would be one certain religion, under the wings of the same guiding and creating entity. The fact that there is a whole range of religions and faiths only indicates that it's an instinctive urge in a human being. Therefore religions were created and originated by the same fertile instinct.

When Moses was gone for forty days and forty nights, the Israelites felt exposed and helpless against the new insufferable situation. As you may know, this resulted in the desperate creation of a new and different faith: the golden calf.

In this state of mind, humanity instinctively and subconsciously feels the need for a strong figure to take responsibility for the events that ensued; a figure that will provide support, a sense of stability, an answer; who will solve problems, assume responsibility, lead us on the right path, take care of our well-being and so on; and who will even think for us. It is like the matriarch female elephant leading the herd on its way to survival. As far as the herd is concerned, the matriarch female elephant is the "wheel" for life. Without her, life will fall apart. It's like driving a vehicle that has no wheels.

Since man knows how to think complex thoughts, even without having anyone to guide or instruct him on how to use his own brain, the "self" develops various beliefs. These originate according to infinite and individual interpretations and world views, against endless situations. These are situations that are forced upon us, like famine, diseases, plagues, wars, injured people, people getting killed, and strong urges not being fulfilled. Weird phenomena, like earthquakes, and volcanic eruption, lead to confusion and wondering. All these cause man to abandon his logical, realistic, and constructive thinking. These beliefs take the place of an absolute and formal religion, developed according to numerous interpretations by infinite, unaware, and individual sensations. Without being aware of it, man steps out of reality. He feels that he is being governed by what he considers to be a legitimate and individual entity. And so he enters his bubble, and from that point on, life is governed by that unshaken belief, for better or worse. In his estimation, his faith is absolute and legitimate.

This causes man to conduct his life by individual beliefs, which coincide with one another. From there on, the path to war, destruction, and great evil is a short one. Paradoxically, these acts and negative results are legitimate in the eyes of the believer. At the same time, other religions lack such legitimacy in his eyes. There's no point in arguing with a believer. It's like trying to describe what colors look like to a person who has been blind since birth. Man perpetuated this paradox, which has resulted in constant frictions for thousands of years, blindly causing him to wreck his own home: the planet.

These frictions are zealously created according to the principle of belonging. That is because each believer is convinced of the absolute legitimacy of his own religion against that which he considers to be an outrageous lack of legitimacy of other religions. Each believer views his religion through different eyes, according to infinite individual interpretations of the infinite "self."

It is the "dark maze" without any road signs.

I should add that belief and religion both "silence logic,"

since belief is the "wheel" in the life of the believer. The phenomenon of imposing your faith upon the other is governed by the dominant sex – the male. Therefore, women, children, and animals are all poor victims and they are, in fact, living and inferior objects: the possession of the man. Despite the irony, even believers are not free of misery, because if it weren't for their misery, there would not have been faiths and religions in the first place. This coercion is done by a subconscious instinct. It's a feeling of hidden jealousy that says: "I stick to the rules, laws, and severe prohibitions and I pay for it by not expressing my freedom, while others have much more fun than me." It's an insufferable feeling causing the man to dominate others to elevate the sensation. This is according to the "quicksand swamp" and the "scales rule."

The question that should have been asked is this: "Why does man force his religion upon the other so zealously?"

We should be aware of these procedures so that we can have a way out of this demonic dance that we are all trapped in.

There are different religions that believe in the same imaginary entity. Each religion was supposedly commanded by that entity, to destroy others while it had ordered the same to the others, going back and forth.

A final example that demonstrates this principle is that of a man living on his own on a deserted island. He doesn't believe in one particular entity because he is disconnected from the civilization's belief system. Say that person falls into a deep hole. By sheer desire for life, he manages to release a cry for help: a cry that supposedly comes from an imaginary entity that doesn't exist in real life. That missing entity is the individual entity of every believer worldwide. It creates an individual reaction that generates infinite interpretations of that same entity.

In conclusion, it's not the brain, but rather the "self" that is using the brain according to infinite individual interpretations, for better or worse.

Therefore, the existence of a man depends upon him alone, and there is no savior to rescue him, like there never was one.

The religious believer is convinced that food and natural resources are given to him as a gift from the entity in which he believes. Therefore, as far as he is concerned, we should continue to bring a lot of children into the world in the name of that entity: all despite the hunger in the world and the clear mismatch between reality and faith. To have a better indication and understanding of natural resources, let's take a trip to the harbor and try to think of where all those massive amounts of products come from, how they came to be there, and what their final destination is.

Another phenomenon that somehow seems to disappear from the media's attention: the crowdedness of ships around the world causes, among other things, the killing of whales that are the last to survive the carnage.

The question to ask is therefore: will this ever end?

In terms of making sacrifices, man has been creating strict rules and he also chooses whom to victimize according to these rules. Why? Because he can. It's not fair.

In the name of the animals: "We animals are crying out: Please, humans, have mercy on us and stop increasing our numbers for your purposes. We also know what suffering is. We sincerely thank you for your understanding and wish you only good."

It would be presumptuous to call and ask societies to stop enforcing their beliefs and fixed norms, which they willingly and passively took upon themselves with their heads bowed down. And yet, we must make use of our rational logic. Indeed, every individual is certain of his beliefs and is fanatical about them. However, the other is also convinced of his own beliefs and is also fanatical about them. Thus, the millionth individual is persuaded of his beliefs and is fanatical about them. Likewise, the billionth individual is convinced of his beliefs and is obsessively fanatical about them. Here is the question which should have been asked: Who is right? The super brain of the man is there, but so far his owner, the "self," has been trapped in an individual, endless, and dark maze. This book is supposed to force man to light up the "dark maze" and start over. It's the only chance to be saved from a tragic ending.

IS THERE A CONNECTION BETWEEN THE SIZE OF THE BRAIN AND EATING MEAT?

Repetitive claims have been made about the fact that the human brain has gotten bigger since man started to eat meat. Proteins accelerated the growth of the brain. If that is the case then why didn't the predators' brain get any bigger? The scientific claim is that the prehistoric man started to cook the meat, which made the digestive system run more smoothly and was the catalyst for the brain. Well then, what about worm eaters and shell fish? Or the spider, sucking the fluid content from its prey? Or the poisonous snake, whose prey is already half digested by venom? To create cooking fire for food, much complex thinking is needed. Like collecting dry wood, bringing it with leaves and dry twigs and using a "lighter." All these actions require complex thinking and other various logistic solutions. Modern man finds it hard to handle these conditions.

Let's conduct an experiment: ask modern people to build a bonfire in the savannah under prehistoric conditions. You'll be surprised to know it's not simple at all.

In that case, what was created first? Was it the size of the brain, created in order to start a fire, or did the lighting of the fire contribute to the size of the brain?
The behavior of animals is based on focused instincts in the present, without any need for complicated interpretations of situations. Complex thinking exists among smart animals like dogs, parrots, monkeys, and dolphins. But the minute prehistoric man started to think in such a complex way as strategic thinking and survival mode, there was no way back, since difficulties in reality required thinking of solutions. A weighty difficulty demands more thinking and

an answer. And from this moment on, man has been using less of his short-term instinct, which he has been gradually replacing with complex cognitive thinking and starting a chain of reactions. But all these events are relevant for the human survival conditions in the savannah. Since there is a big brain and complex thinking, but no one to guide thoughts and judgments in the brain, the "self" thinks independently by its infinite individual interpretation, through the non-understanding brain that passes no judgment. The results speak for themselves: some manifest in sliding into different religions and beliefs, thus causing the deadly collision between the two.

The results are therefore: helplessness, despair, mania, violence, frustrations, revenge, and mental illnesses. When the results are negative, the believer is more deeply "rooted" in his belief, innocently thinking that the gods are angry with him and should therefore be reconciled.

Moreover, a man was blessed with a large brain and the ability for complex thinking. But he gives up on this bonus (complex thinking) without being aware of it, and instead uses his brain only partially, since the reality and its results aren't satisfying, and are even disappointing and burdening. He'd rather not be aware and bury his head in the sand, saying different things like: "It will be okay" or "It's not up to us," or "The 'entity' takes care of it, etc."

And is the human brain too big? Does size matter?

If so, why is there a big nose? Is it a nose that smells better?

The fly has the ability to smell better than the human being. If so, where is the size proportion here?

The human brain is large regardless of its quality. **The human brain is large because man is constantly full of thoughts**, whether they are positive or negative, right or wrong, manifesting themselves as expressions of body language, and aiming toward survival strategies by thinking.

Another reason is that **there is no creator of man regulating the size of the brain and maintaining an adequate and desired size.**
Theoretically, some men can be born with smaller brains than others, but with the same quality, as a small child doesn't have any less of a brain than an adult has, just because of the difference in the brain size. Since man has no entity to guide his thought, he does so on his own, by the "self." Still, he only knows how to use his brain, whether for positive or negative purposes, in a wasted and sloppy manner.

Man is born on a beautiful planet, and it is, in fact, a great privilege since, as far as we are concerned, there is no vibrant life on any of the planets surrounding us. So instead of appreciating the extraordinary, man destroys his magnificent home and even does so with great pleasure. For example, on one hand, man has created a state-of-the-art gun, requiring great skill and complex thinking. On the other hand, he uses it to slaughter the spectacular and marvelous rhinos, for his own sick purposes. Moreover, anytime he manages to kill this beautiful animal from a safe distance, he creates a V with his fingers, and his smile shows great pride for this "heroic" deed.

In conclusion, thinking requires **mental** energy. That's why the brain is operated by its owner and wastes energy, unnecessarily, since the body doesn't work on the basis of needs. For example, when we have a sense of frustration, or of chronic fatigue and weakness, the energy levels in the brain drop according to these sensations. At the same time, the owner of the brain – the "self"– often compensates by inducing complementing energy from high energy foods (carbons, candy, etc).. It's a sloppy and senseless waste, but that's how the body works. It's an instinctive reaction, yearning for life and energy. That is the need to rise upwards according to the "quicksand swamp."

THE ANTS

Ants have rich and full lives, filled with urban, vibrant, and non-homogenous activity. They run their lives with a microscopic brain and operate survival techniques that are more complex than those of any of the larger animals, such as whales. They do so quite astoundingly: on one of the National Geographic shows, ants were shown arriving at a place that could not be crossed. However, without much contemplation, they created a bridge using their own bodies, in order to enable the other ants to pass above them, so they could move ahead. Also, when the queen was caught on a leaf, floating away in the flood, the rest of the ants protected her by creating a shade. They even cooled her from the heat.

Another astounding move is the kidnapping of larvae. The ants invade another nest, fight the "soldiers," and prevail. After that, they snatch the eggs from the enemy, and raise them in their nest, for them to assist in the daily chores in the future. The irony is that the larvae that have developed in the new habitat aren't aware of the fact that they are kidnapped slaves. They are sealed from the moment they are born by the adoptive family. The larvae feel obligated, and belonging to the place of birth and family. That means that unlike human slavery, they don't feel humiliated and downtrodden.

I strongly encourage watching this program, not strictly as spectators, but rather in a way that we can all see ourselves as a part of them in each single move they make. That is, we should see ourselves joining them in building a mound; helping them construct an intricate, hollow, and sterile construction with ventilation; using all the logistics that are involved and living our daily lives as a single ant among many others.

This experience will highly contribute to the understanding of the brain and its size.

Optimism

Optimism is a positive and rejuvenating quality. Yet blind optimism with regards to reality borders on naivety at best, or stupidity in the more dangerous case, and can also end up in disaster. This quality has caused many tragedies and disasters in the past, as men with blind optimism were always oblivious to the risks brewing around them. Therefore, pure optimism, without use of the senses and logic, doesn't contribute to survival in any way.

Healthy optimism is when a man is convinced that he is

overcoming what is twisted by realistic actions. Therefore, leaders should possess this quality. **Many disasters and tragedies have occurred because of diseased leaders.**

GAINING WEIGHT

There are no external entities but us worrying about our design, like: designing us to be thin and efficient. It is the individual "self" that determines and designs our existence without us being aware of it. A fat man feels "fat." His character is "fat." His habits and moods are "fat." His body language is "fat" and he visually sees himself as "fat." Therefore, the body operates and behaves in accordance with that "fat" perception. That's because the brain is not an understanding organ that can lead us on the right path. It doesn't fix our mistakes and doesn't reveal the reasons for the weight gain.

Gaining weight begins somewhere. It is the loss of the individual "self" that lets the body go and relies on its owner instead of being designed by him. It is a loose and fat "self" that governs the body and plays tricks on it, instead of the designing "self" that should have guided the body. The fat then becomes distress and an embarrassment. It's then that the person starts taking care of himself with physical training and low calorie food, but with unsatisfying results, since he had chosen the wrong path. That's where the frustration and snowballing effects originate. He is already convinced that he is fat and a fat sensation is followed by an even fatter sensation. At a certain point, he reaches a dead end. He deserts his battles with windmills, claiming, "This is who I am and there's nothing I can do about it, so I'll eat as much as I want, and gorge on whatever I want, since there's no way out anyway." And the results are known in advance.

In fact, there's no direct relation between the amount of food consumed and gaining weight. Despite that, overeating could cause indirect weight gain, due to the "gluttonous" and "gorging" nature of a fat person.

As for children, the motherly instinct says: "Eat up." "You must eat." "You must leave an empty plate." And "delicacies and snacks aren't food." This is unconscious and even an innocent poisoning. This phenomenon only exists among humans because the motherly instinct (that follows the belief that food is highly important) overcomes logic. We should be aware that this phenomenon is sealed in the memory, and could cause obesity or anorexia. This is done out of aversion and defiance toward the repulsive and rejected time of childhood. It's a kind of (subconscious) "I accuse" expression.

What to do?

Recommendations for handling weight gain

The first step is to purchase clothes in our skinny goal size, which we will be motivated to reach in the near future. This is to set an indisputable fact, so as not to give way to skepticism and lack of determination.

We should deny the body as it is. We should also step out of it using our vivid and diverting imagination, and abandon it. The second step is to design a new image of a thin and athletic figure using our great imagination, and "fitting ourselves" into it, just like trying on new clothes.

The most important step is to feel the new, stretched, lean and athletic body and sustain it in a continuous manner.

It would be better if we turned all our mirrors around so that we aren't forced to stare at this body that we are denying.

Next, we must change our body language from loose to efficient and quick.

We should learn to avoid supporting our stomachs while putting on shoes and tying our laces, and not support our knees using our elbows.

In fact, we ought to put our shoes on with one leg in the air. Yes, we should do that with our thin bodies, lean hips, and empty stomachs in one instinctive movement.

Don't lean on the chair's backrest in a calm and peaceful manner.

Don't look for comfort or relaxation during the time you sit, sleep, or even in general.

Be alert even while you are sitting.

Such instructions would change the "fat" body language into a "thin" and light one.

We should all eat, behave, and think thin. The more this insight is understood, the more vivid the thin sensation becomes. Meaning, I am on my way; the past that I couldn't let go of is already behind me.

Physical exercise followed by a "fat" sensation won't bring any results. We must, therefore, feel thin in all our organs.

An important comment: imagination itself, without the added sensation, won't manifest itself. This insight is generally relevant.

An exercise, for example

Spread marbles around the house. Collect them in a "thin" way, using thin body language with thin and high-quality bending. Meaning, a body that doesn't sigh, seek to lean on objects or hold on to anything; a body that isn't slouched over or hunched and isn't just dragged along. The body is a copy of its owners, created by their sensations.

It would be useful to carry on with the plan until we get positive comments from people we know, like: "You have lost some weight."

From this point on, the fat sensation will alter into a thin sensation. It's the moment where the "self" is convinced that it is thin. It's a vivid and more real step than the one at the beginning of the plan. The sensation is the determining factor. It's a feeling of transcendence that says: "Yes! This is real. I have discovered the principle."

The moment the treatment is proven successful, the "thin" sensation emerges (which is the most important), and the principle is thereby understood.

There goes a chain of positive reactions and the sky is the limit.

Comment: it is advised not to share this plan with anyone, especially your partner, since they will instinctively mock and make fun of it. There are many reasons for that, originating from the depths of the subconscious, like the envious sensation that says: "She will be beautiful and sexy whereas I will become uglier." (according to the "scales rule"), posing a threat to the relationship and a fear of cheating.

ABOUT FOOD

Half the world is starving, while the other half is overweight. The starving people die not just because of illness, as you might expect, but...right! Because of starvation. The overweight half, on the other hand, live a life revolving around food, even though they have plenty of it. It's another proof that there is no such thing as "healthy" food. "Healthy" food is an illusion created by man. It's a result of man's helplessness and his lacking the solutions for problems, which in turn leads humanity to holding on to even more unfounded belief.

As for food: we should eat a little when we sense hunger, and by the beginning of satiation fly away from the dinner table like a bird.

Here's an important insight for those who complain that their fridge is empty (and to rich squandering people as well):

The phenomenon of having cooked meals, while sitting around the dinner table, is an invention of mankind. In fact, **biologically speaking, a human being can exist on strictly dry bread and lukewarm water or on wheat seeds**. It's definitely sufficient. The conscious "self" (including that of animals) is the one that longs for diversity in tasty food and a "tasty" life in general to subconsciously improve the sensation). Did you know that a large variety of food is actually made out of the same ingredients as bread (wheat)? Wheat has many advantages: it grows in rain water, it does not rob sweet valuable water sources, and it does not need all kinds of sprays and chemicals (which cause a lot of damage). Therefore, **it is** the preferable food for man.

Food should be a minor and trivial consideration in our daily lives, unlike people whose lives revolve around food, convenience, comfortable sofas and armchairs, and a general "fat" atmosphere. We should become aware that a fat society encourages weight gain by subconsciously imitating the other.

We should, therefore, resent such revolting lifestyles and feel contempt toward them. I remind you not to forget the thin new body you just put on yourself. It's a new and thin costume the new "self" has put on, and which should be kept forever and with great zeal.

A comment: if a fat person weighs himself every day, and mumbles sick and unserious sentences like, "I'll tryyy," or "We'll see if that helpssss," it's better not to begin the plan for change quite yet, because this man has yet to understand the principle of our creation.

As for parents who cause the obesity of their children: they are in an inferior position and supposedly resolve their own pain using food. The reasons are varied and subconscious.

Obesity is an immoral and repulsive offense that damages morale. Parents who make their children obese are nothing short of moral offenders.

As for compulsive eaters whose lives focus on cooking and eating high-quality meals: leave some natural resources for your offspring – did you ever think about that? Don't leave them without any!

As for liposuction surgery: this is one of the more common biological ironies. After all, fat is just that – fat. We can easily get rid of it while the obesity principle is internalized.

WOMEN

When such horrid cases as breast tumors occur, women should, despite everything, nurture their chest and get in touch with their bodies. Don't become compliant with the disease. Don't wait for miracles to happen, because there is no such entity that will cause miracles. Women suffering from such diseases should imaginatively divert the phenomenon and change it into perfectly healthy breasts. Those women should walk upright with their chins up and their chests filled with pride. Don't let go of this feeling for a second. It is recommended to walk with a fast pace and an erect chest, because this body language conveys the sense of determination that is invaluable in this case. They should take care of the "self": nurture and empower it. Skepticism shows a lack of understanding of our entire creation. The key for the success of this plan is sensation: taking pride in the beautiful and natural chest. This is done by imagining empty spaces (according to sensory and divertive imagination).

Urinary tract infections or similar diseases aren't caused by an infection or lack of hygiene, bacteria, or fungus. It's time to stop this misguided perception. These illnesses are caused by the "quicksand swamp," which leads to the collapse of the overall physical system (see **"Obsession"** episode).

The fungus, for instance, develops once the body's owner is in a deteriorated state (such as being "murky," "rusty," and hurting). As you probably know by now, when a certain phenomenon is frustrating for us, focusing on it will only lead to extensive frustration. And the frustration will enhance the disease which, in turn, will accelerate the frustration and so on and so forth: the snowball increases.

We should start our lives all over again. Sit down with a good psychologist, discovering the essential hardships in life that are inducing the feelings of inferiority. Start living in a new and different way than in the past, in order to see the world and life through brighter eyes.

All this is in addition to the "resistance." The "resistance" supposedly says: "There's a lethal acid eating up diseases within me."

In conclusion, that is the "*Ultimate I am my own maker*" feeling these illnesses and his body is based on his endless interpretations about life.

DIABETES

Now that the principle of our creation has already been internalized, the principle of illnesses should be comprehensible as well. Diabetes is one of the diseases that characterizes modern man. If we compare modern man to wild animals that exist under strict and relentless conditions, the principles of the diseases and their causes should become clearer. The cause of diabetes is the polarized change from nature's rugged life to the looseness prevalent in the life of modern man. Modern man has become a depleted, dependent, passive being, living with the knowledge that his body isn't his responsibility and he's not even attached to it. A body that is neglected by its owner becomes distorted.

Therefore, we should become aware of the above-mentioned factors, and part with such "sickly" habits and behaviors. We should also know that these diseases cost the public a fortune and even damage morale. We ought to restore the body to its determining "*Ultimate I am my own maker*" state, by being aware of the lifestyle of wild animals. Animals don't rely on anything but themselves, and we too must realize that no "magic potion" can solve our problems for us. There is no food that guarantees our health. There's no particular entity other than "I am my own maker," which is the ultimate owner of every living creature. We should step out of the dysfunctional framework that characterized modern man and learn to think like animals in nature.

Parents should stop sending "sickly" messages to their children (which we have already discussed) like: "It's healthy"; "It makes you stronger"; or a sickly attitude like: "Sweeeetie," or "hoooney pie." We should treat our children as the survivors they are and prepare them for the struggle of adapting to the circumstances of reality. Reality is rugged. Don't deceive your child by any other illusion. The child is supposed to grow, blend in, and assimilate according to reality.

This is a question we've yet to address properly: "What is the connection between high blood sugar levels and diseases, including the loss of organs?" It is again the "individual self" which is passive about his body, unaware of it and detached from it.

Unlike diabetes, lack of blood sugar is a sweet-less sensation in life and lack of energy.

Treating young people whose bodies have become a burden to them would go as follows: take a helicopter and parachute, give them a small shove, and allow them to skydive into the heart of the jungle – should I say more?

HOARSENESS

Our voice becomes hoarse when we "feel" hoarse and mentally cry out for help. This usually occurs during a downward spiral, when some situations in life cause us to feel the following: a strong urge of expression, saying what is weighing on our hearts, and fulfilling it by talking, protesting, or singing out. These sensations exist when there are barriers and fears, usually of ears that don't listen to us; and it is in those moments that we begin thwarting those passions within us that are yearning to be expressed.

It's a kind of "rusty," exhausted, and "lacking" sensation

(like a "short circuit") that is crying out for expression and conduction. Since the body operates according to the sensations of the "self," the vocal chords act and operate by that same worn-out feeling. Not because there is an actual exhaustion, but rather because there is no contact that would cause biological exhaustion of the vocal chords. After all, hoarseness could occur without speaking – even a mute person can suffer from hoarseness.

Recommendation

We should moderate those burning sensations that are yearning for expression. **Extreme phenomena such as these damage the body, according to the extreme sensations of its owner.** We should express ourselves and flow freely, with self-confidence; thus removing all threats that collide with the same sensations. We should order our "self," telling it: "I follow my own pace, without strain."

LEFT-HANDED OR RIGHT-HANDED?

Individual information lies in our subconscious. The consciousness is like a spot scanner out of which the conscious "self" pulls information. Therefore, using the body, or for

our purposes, using one hand, is a more efficient and focused use than using both hands simultaneously.

We are born with diverse qualities. These qualities aren't
carved in stone and can surely be altered. For example, a
person who is born left-handed and was forced in school
to become right-handed will indeed become right-handed,
but only if he is truly convinced of the benefits of it and
was appropriately compensated. Forcing a change without
being internally convinced of it won't generate a true
change. This coercion could cause disturbances.

So is the case for every trait that we are born with. As
time passes, the particular quality becomes more and more
rooted within us, creating fixation. Therefore, when similar qualities run in the family, or even similar illnesses,
they might not be hereditary. There's a possibility that it's
a family condition, acquired by behavior passed from one
to another, when there is the same atmosphere within the
family, alongside such familial traits as similar speech style
and behavior – everybody with the same habits and body
language.

As we have seen, habits grow over time, during generations.
And they are affected by similar genetic qualities, one way
or the other. That's why genetic diversity is preferable

to marriage within the family. It will serve to diversify character. Because a certain character is exposed to certain illnesses – and that is the individual structure of the "self."

Evolution isn't responsible for genetic diversity, neither among animals nor human beings. **Evolution isn't a conscious and comprehensible entity. There is no guiding hand to ensure genetic diversity.** But in an instinctive way, an individual is sexually attracted to a character that is distant, qualitative, attractive, new, different, intriguing, and more interesting than the people he grew up with and whom he had grown accustomed to, so they no longer intrigue him. The individual doesn't operate according to some calculated need for genetically diversifying his offspring, but rather as an instinct, seeing as animals and the prehistoric humans knew nothing of genetics. For instance, when a brother and sister who were separated at birth meet randomly at age eighteen, they don't know they're biologically related and might be sexually attracted to one another. The brain doesn't ensure genetic diversity, nor does evolution. Sexuality is a very personal sensation. It is (unknowingly) directed toward a new, distant, and charming other.

The instinct in this example has already created diversity without being aware of it. Let's look at another example – obese parents. They might pass on "fat" qualities to their son who might be born thin. This child might grow up and imitate his parents.

THE EXTINCTION OF DOMESTICATED ANIMALS

The principles of our creation and that of the animals are the same. For instance, during a time of distress and anxiety forced by humans, the animals' fur becomes worn-out, stiff, and rigid, in accordance with the grim sensation of the "*infinite self.*"

I present to you a rhetorical question: a bird, a parrot, for instance, is trapped inside a small, permanent cage all its life. Is this acceptable to you? Do the depression and anxiety of a parrot originate from a lack of serotonin in its body?

Wild animals survive according to the laws of nature, but man doesn't obey these rules, and therefore causes the distortion and extinction of the animals. The thoroughbred horses, for instance, are getting weaker and weaker with each passing generation that creates a mutation in those horses, leading them to their inevitable demise. The facts speak for themselves: they are too prone to illnesses; they have fragile souls, weak bones that are breaking; and their hooves dissolve as their quality deteriorates. That's because this species is brought up in laboratory conditions. The irony of the matter is that the mutations actually improve because of the existence of predators, threats, and social rivalry among themselves in the wild.

This sort of spurt, therefore, requires improvement, strength, and a positive mutation. If it weren't for these principles, there wouldn't have been an evolutionary development and there wouldn't be any animals alive today. Meaning we would remain microscopic creatures. But domestication came along and created a distortion of the laws of nature and their results.

For instance, a foal is born into conditions that don't resemble the wild one bit. He grows up on bedding, in a crowded stable. When he's in the meadow, he stands on flat, homogenous grass. His life is conducted in a boring, empty, risk-free, meaningless routine lacking any resistance urges. From a young age, foals are fitted for horseshoes, which they wear for the rest of their lives; thus, becoming fragile and sterile creatures. As the generations pass, history is imprinting its genetic characteristics in the nature of the creature and the results are before you – the horses become feeble creatures; all because of the personal, self-benefiting interests of the horse-breeding industry.

We should take certain measures in order to improve their lives. For instance, fill the meadow with obstacles that are similar to those in the wild like rocks, pits, wooden logs and no soft grass. Obviously the foal should grow up with its peers, **to which it is accustomed**, and be provided with the most healthy and vibrant life possible. However, we must toughen up the horse moderately by taking its mother on short trips every now and then, thus, following the foal's urge and desire for independence. The foal must then join its mother in the open field only when it strives to be reunited with her by throwing tantrums, releasing energy, studying its surroundings, and rubbing itself against a stony, heterogeneous soil and feeling satisfied doing so. That's how it learns the rules of resistance and becomes stronger.

The basis of resistance systems is strengthening the animal, making it stronger and more resilient early on in life -- when the foal becomes familiar with the obstacles of nature. To help the horse adjust to nature, we must place its feet on rocks, make it skip over various obstacles and overcome them, and teach it to resist sliding on the mud. In this way it would train its body for inconvenience, resistance, and proper use. This is the way for a horse to become stronger and learn, unknowingly, that this is how nature is constructed. The foal will carry these lessons its entire life, since memory is not erased. As for the biological parents, we should keep on riding them without horseshoes, to improve the genetics by positively mutating the traits of a riding horse.

As we do this, we will probably learn that the foal is more immune to disease, and is even physically and mentally stronger. Routine and lack of enthusiasm in life cause various illnesses.

I personally request that all animal breeders stop treating animals as mere biological machines.

Please read the animals' petition, which I have made on their behalf:

"We domesticated animals don't like boredom, routine, and various injections. We like to smell and grope nature, study our bodies and use them. Let us be free, take us on trips, diversify our lives, give us good company and space, and we will prove ourselves efficient. You'd better change your attitude toward us. It's about time we emphasized the fact that we have been your slaves and victims for thousands of years, in case you haven't noticed.

You should ask for forgiveness by doing some positive deeds."

As for injections, let me tell you about a disgusting phenomenon that would allow you to judge the difference between stupidity and absurdity, and sanity:

I have mentioned throughout this book that efficiency manifests itself by the condition of the "self." Therefore, as much urge as possible (self-esteem, enthusiasm, etc). and as little humiliation, excitement, and fear as possible. That way the body reacts more efficiently.

There is an absurd tradition, according to which, on the way to the race, the horse (Flash) is kept in a burning hot, closed, and scary trailer, (sometimes for a whole day, due to technical problems) and given various injections, highly recommended by science and biology (such as a diuretic, B12 vitamin, and so on). May I remind you, the horse is under a strict regime of induced thirst, in addition to the use of diuretics, which dehydrate and cause thirst. The horse is naturally scared, since it doesn't know why it's being treated this way. As far as the horse is concerned, a punishment follows a crime (like kicking or biting). The horse obviously refuses, and goes on a rampage, because it is overwhelmed by a sense of fear and terror regarding the situation it was pushed into. Not only does the horse believe it is being punished, but it is physically tied so it cannot move. If it does manage to move, its upper lip is tied in a "twitch," which is a rope, tightened in a circular fashion around the lip to aggressively and forcibly strangle the horse. And if that doesn't work (as a result of enhanced fear), you add another twitch to its ear, to force it totally. All this is done in a blind and idiotic way, as they treat the horse like a wooden horse.

As for diuretics and urine-inducing substances, these reduce the pressure on the biological systems during the race, according to human understanding. This assumption seems to have disabled the rational logic of the horse trainers, since no one thought of declaring this an inappropriate practice, **particularly** under the impossible circumstances of the hot summer conditions, causing the skin to completely dehydrate. Bleeding lungs are a result of a burning sensation in the trachea and lungs, when the urge to implement the task is missing. The diuretic does not solve the bleeding lungs phenomenon.

By the way, it is possible to teach a horse how to pee. Each time it pees, we should stroke its nose and say, over and over again: "pee, pee, pee." Moments before the race, take it aside, so it will relax from the excitement, and use this trick to help the horse empty its bladder.

It's a shame that man doesn't make these sorts of experiments on himself. It has never been unequivocally proven that substances only have a positive influence, like fuel supplements. These substances supposedly have a positive effect in a lab, but they only cause the subject to have wrong interpretations of his true sensations. That's the difference between sanity and total stupidity. In addition, despite the fact that the injection of the substances doesn't help performances, and the horse is backing down due to recurring trauma and fear, the same tradition goes by blindly. Because belief is belief and is undisputed ("pumping" with steroids causes apathy).

I again repeat and declare: an animal doesn't forget its past, especially traumas – just like us humans. You probably know the phenomenon of chain of reactions. Due to the mental and humiliating previous burn, the horse refuses to get up on the trailer the next time, since naturally he's not willing to go back to hell. Again, the expected begins. Meaning, using force! Putting him forcibly into the trailer, using different kinds of twitches, and this will repeat itself until the collapse.

And that's not all: on the exit from the sprint facility, the horse's rear side is whipped to accelerate his sprint. All this is done in a blind manner. His situation is already inferior, lacking in energy, urge, and a desire to run. They are, in fact, beating an already beaten horse. It's truly a tragedy.

The result that follows is a hurt and forced horse. He opens his mouth during the race as a result of the stress he is under. The common "solution" is to close his mouth using a tight rope. That is, without looking for the actual reason as to why he's opening his mouth. It's like a father putting a band aid on the lips of his son when he's crying. It's a tragedy. What levels of cruelty can humanity reach? Or more precisely: what levels of cruelty can MEN reach in their pursuit of their lost honor! Twisted honor. When someone gives him a constructive comment or recommendation, his twisted honor is shaken, causing him to instinctively and immediately do the complete opposite (see "**My Story**" episode).

All the attention of those involved with animals is given merely to the few and privileged winners. Sooner or later those "objects" that "broke," or "went out of order" or the unsuccessful that "became obsolete" are sold in an auction to whoever wishes. These "objects" go through hell to their death – a fact that is hidden from our awareness time after time. Again: this is "the dark maze" among man.

The yak is a Tibetan beast. The Tibetans have used it for hundreds of years for excruciating journeys and exhausting agricultural labor. They drink its milk and make cheeses from it. They eat its flesh, sew clothes from its hide, and use its droppings to warm themselves and make fire. In addition to this, they tie them up, groaning, in no man's land, in harsh winter conditions. As far as they are concerned, that is the role and the calling of this "machine." Life is exhausting, according to the "quicksand swamp." We and animals alike cry out for positive and purposeful changes, and yearn to improve our sensation as we keep on grasping onto life. In phases of exhaustion, the Tibetans turn the yak over on its back forcefully, pushing a kilogram of pure liquid salt into its jaw, supposedly to renew its energies due to loss of salts and electrolytes. Some of it drips into its eyes...stupidity. That's how they show their appreciation.

For those of you travelling to Tibet – please pass the Tibetans the next message: treat your animals as you treat yourself. Add the appropriate amount of salt to their food, divide it into small portions, and serve often. Alternatively, you can feed them with floury candy, which contains salt as well. This bonus will change the sensation from a bland to a positive one, and will renew the energies. Is this a complicated insight?

Adding the right amount of salt provides a positive sense

of absorbance which contradicts the bland sensation. These surpluses cause nausea and poisoning. A spoon of pure salt causes nausea and poisoning. A spoon of salt dispersed into food doesn't cause those symptoms.

COPYING SENSATIONS

Copying sensations means imitating the sensations of others (in a subconscious way). Like crying, for example; when a man cries for some reason, a different individual witnessing his cry might get teary eyed as well, or even begin to cry himself, because he honestly and sincerely empathizes with the other's sorrow; he relates to the other's sensation of pain and in a way, merges with the other's sensations.

Another example is yawning. When a person yawns, another individual might join his yawns, if he senses in his imagination the empty, boring, exhausting sensation, together with the "no action" atmosphere that the yawning person feels and merges with his sensation. All of this is subconscious, of course.

Yet another example is laughter. When a person laughs for

some reason, another individual might join in his laughter, if he feels the "laughing" sensation of his friend and merges with it.

It is the infinite "self" that feels, imagines, relates, partakes, imitates and merges, with the other "self."

All these are copied into sensations. The possibilities are endless, and some of them are as follows: copying or reflecting another's character, opinions, beliefs, habits, norms, body language, diseases, and mental disturbances.
As a matter of fact, most of our lives we waste a lot of energy (without our awareness). For example, a mother gets exhausted because her son goes down the slope and fails. She tries very hard to draw him from the deep, back to the top.
All the different diverse and ridiculous theories are irrelevant.

ADDICTIONS

The body doesn't get addicted to various substances, because it's not the one calling the shots. Rather, it is the body's owner, the "self," that becomes addicted to a certain sensation following an experiment with a substance, which is usually accompanied by a habit that diverts the person from his sense of reality. The causes of addictions are a mixture of sensations: peer pressure, boredom, imitating others, need for attention, escaping from a disappointing reality, a sense of defiance in general against a disappointing life and so on.

On the contrary, there are many people with similar difficulties who don't automatically comply with social obligations. Unlike those people, the characteristic of the addictive type is a loose "self." A boy who is caught in the web of addiction is a boy who has no "steering wheel" to hold onto in his life; or he might have one, but doesn't know how to "steer" it. The addiction has a beginning, and he is not yet addicted after the first several experiences. But it is a direct route to the abyss.

It is the result of poor education leading to the inability to differentiate between right and wrong, and between what is and is not allowed. (This comment is directed toward parents and the educational system). There are many traps in our lives. A true survivor should dodge them.

The inferior type will get caught in these traps. We should convey these messages in school, in order for each boy and girl to be aware of their actions, and avoid the embarrassment that is the trap of addiction.

A boy who feels inferior to society will see himself through the eyes of others. Because of that, his actions will be influenced and determined by others. Let's say he takes a cigarette so as to impress and look sophisticated. As we all know, the cigarette will eventually burn out in a few moments, and then he will have to smoke another one so he will not be perceived as unimpressive and boring. Once this is over, he adds alcohol as well. As can be expected, he will at some point run out of drinks and will be uncontrollably tempted to take another…and so on.

No matter what issue we examine, we will find the "quicksand swamp" law in each one. Compulsive gambling, for instance; the compulsive gambler finds himself chasing success and victory to elevate the sensation. It should be pointed out that success, whose results are known in advance, is boring and unchallenging. On the other hand, obsessively chasing victory or success, when the results are not predetermined, is like being in a dizzy spell. It's a tempting trap luring its victims in, yet it continues to spin endlessly, because you never know what the next move will be. Losing a bet will cause a sense of humiliation and crashing that would lead to another attempt at the card table in the hopes of succeeding.

And success, as you may well know, lures a person for another attempt at success, repeatedly and endlessly.

An addiction to the act of smoking cigarettes comes out of habit. The dissatisfaction of a habit, as you know, is frustrating. The frustration mixes your senses, personality, and energies. Therefore, there is a duty of satisfying the addicted "self."

The cigarette smoke doesn't contain any energetic substance; it is the "self" that is now satisfied and balanced.

Cigarettes don't contain any addictive substance in their smoke – it's the "self" getting addicted to smoking.

The cigarette smoke doesn't contain any substance that causes ulcers and has no direct connection to it. A man who is prone to ulcers is a man who is generally sensitive to certain (spicy) foods, burning smells, grating sounds, and so on. Due to his condition, he struggles for a pleasant, clean, sweet, and peacceful and atmosphere, which he can view with rose-colored glasses. Therefore, the smoke of the cigarette, which contradicts all this, adds to his already burning sensation. Moreover, he is coerced to continue poisoning his body, which is already burnt out anyway, with the smoke. Thus, the connection is there, but it is indirect and not biological.

The body possesses an ability to clean the soot of the cigarettes, but this is not done in an obvious and automatic manner. The poison cleansing is done by the body's owner – the vibrant and alert "self" that is aware of his body and using it.

The cigarette smoke affects the skin badly in an indirect way, in accordance with the "self" who sees and feels the world as black, dark, foggy and choking.

It's not the brain that is addicted, but rather the infinite "self." A theoretical example: let's say that in a moment of longing for the substance, the addict is in some kind of extreme situation, like a fierce hurricane wreaking havoc on its path; all his longing will disperse. This is because of the situation that drives him into other areas. Suddenly the brain supposedly forgets that it is addicted.

Children can also be addicts. For instance, a child could be addicted to having hot chocolate in the mornings. Animals can also be addicted, say to a tasty mix of hay in the morning, noon, and night. But it doesn't mean that the hot chocolate or the hay contain within themselves any addictive substances. Another example: when a child is addicted to a computer. It does not mean that the computer includes addictive substances. Last example: when a boy is addicted to his girlfriend. It does not mean that his girlfriend is addictive. These are merely addictive behavioral patterns.

There is no food that cleanses toxins (that can only work in a lab). In order to step out of the addiction, the addict must learn the reasons for it (the above-mentioned) to help and prevent frustration. He should change his social environment and atmosphere, his clothes, his walking and talking style, and any previous habits; and he should occupy himself responsibly and constructively. He should look back and treat that period as a repulsive and wasted tragedy, and cover the past with a hermetic lid, transitioning into a new era. The most important thing is that the new, proud, and authoritative "self," unequivocally determines the courses of our lives, like: "yes" means yes and "no" means no. We should think like this: "I'm my own tough and determining commander."
It's possible. Good luck.

The moral is: Hey, boy. It's easy to sink into the depths – be careful. If you're trying to impress others by doing more and more masochistic deeds, you are not a man –

you're just trying to be. Meaning, you don't know how to "keep your head above water." You just quiver and waste energies, which makes you inferior and an imitator of others. Use your sharp senses, because they exist. To every mean-spirited temptation toward the trap, such as: "drink alcohol," "take drugs," "drive faster," "don't be a coward" and the likes, answer back with opposition, ridicule, self-control and complete self-confidence: **"My mother does not allow me."** This way you can be saved from the "quicksand swamp."

ALZHEIMER'S DISEASE

Alzheimer's disease is an occurrence. The occurrence, as you know, has ups and downs. Each individual experiences ups and downs in his life according to the "quicksand swamp." In these situations there's a decrease in the quality of cognitive thinking. Alzheimer's disease comes in stages – it is the retreating and sinking "self." It's a sensation that says: "Life repeats itself. It is less interesting and more 'gray.'" In this type of situation, the consciousness is less relevant, and it's harder to draw information from the subconscious. The brain conducts itself with glorious accuracy, based on these processes. And that is how Alzheimer's develops.

It is the "self's" collection of infinite and individual information. This information is set in a subconscious way, according to individual time sensations and time gaps of events; and also according to our interpretation of the moments when we "sink" into confusion, fatigue, senses of loss and dizziness. The order of the received information also becomes confused according to the state of the "self."

That's the reason that there are "sickly" situations like: a man who is reliving his past so vividly that he is convinced that he is there. This situation is legitimate as far as he's concerned, but it coincides with reality. This leads to more confusion, anxiety, and dizziness. The irresponsible brain doesn't maintain the right order of situations or lack thereof (see "**Concept of Time**" episode).

But we shouldn't blame the brain, for it is the sunken "self" that has lost itself.

Recommendations for all cerebral disorders:

A person should fill his life with rewarding deeds, like volunteering and contributing. In the sports field, it is recommended to play table tennis or badminton (not at home, but at competitive sports classes), because this amazing game requires the maximum mental and physical coordination.

There's no potion or remedy to force us to take interest in our lives and love them.

BELONGING

Every living creature is born with an instinct of belonging. For example, child A is jealously keeping his toys, possessions, and friends to himself. The moment child B plays with the toys child A is no longer occupied with, he immediately feels a possessive jealousy, feeling as though, "I am more deserving of these things." In his perception, child B is having more fun than he is and with *his* toys, nonetheless. This example is a demonstration of how each individual strives to be "more" and "above" others. It's the survival instinct operating in sync with the "quicksand swamp."

Belonging is a diverse and individual sensation. We strive to belong to a family, to friends, neighbors, a group, a city, a town, a tribe, a religion, a territory or a country.

The religious believer lives his life thinking that nothing is up to us, and that the course of life is predetermined and written by the entity he believes in. The other conducts his life with a baffling innocence, which causes him to say such things to himself as: "It will be all right. People aren't bad. It's all heaven around me," while filtering and ignoring the horrifying occurrences around him. This has brought great tragedies throughout human history.

On the basis of the above assumptions, it becomes clear that such a task as governing a state, for example, requires an understanding of the laws of nature, to prevent frictions and tribal combats. We should keep in mind that the human being could be very dangerous to his fellow men, to whom he feels no sense of belonging. After that point, he removes himself from his neighborhood, religion, and group, particularly when the threatening group has more births and expands. The taboo against discussing demographic stability is even more so a terrible, malicious and clever crime. And we shouldn't believe the, "It will be all right" sensation, for we all strive for acceptance and company.

Some people identify with the opposing team. This in itself is legitimate and should not cause a problem. But tragedy occurs when people fulfill their passions blindly, without any consideration of the results, whether it borders on harming the team they belong to or the team they identify with. Meaning, the chaos of today is a result of the chaos they've created in the past. This occurrence is also legitimate, but comes from innocence bordering on absolute stupidity. These moods are subconsciously fed by masochism or overcompensating for sadism. Some leaders do legislate positive laws, like hunting prohibitions or prohibition on human trafficking (where women and children are the main victims); penalties for attacking the elderly or abusing animals; and of course prohibiting violence against children (a result of a subconscious compensation for sadism). But on the other hand, they are still lenient toward evil and treat it with understanding, and so the tragedy continues. This paragraph is intended for people who, as it appears, want justice. It's a blind and subconscious hoax, concealed by the mask of justice. Some may get rewards, but innocence conceals the fact that these rewards are given by the same organized cult.

We are talking about some people who fiercely defend evil and, on the other hand, trample the victim and humiliate him. The reasons for this phenomenon are diverse and subconscious, but the purpose underneath it all is to sweeten a bitter heart by degrading the other, causing his heart to become just as bitter; thus, satisfying the bitter-hearted person according to his twisted interpretation.

As long as this conduct goes on, and people like that continue to pull the strings, evil will remain legitimate, and the people "belonging" in its group will go on being humiliated and defeated. If these cults persist and have an impact, there'll be no way out of a dark future.

In conclusion, a man doesn't manage his life "the way it should be," like: for the sake of his family, future, state and home, or even the planet. Rather, he manages his life based on his beliefs, religions, holy places, politics, hatred, envy, resentment, twisted self-dignity, etc. We must be aware that in addition to population increase and its distorted heterogeneity, every phenomenon is contagious and spreads and causes cumulative damages. This principle is identical in the animal world.

Bear in mind that man does not necessarily behave or think rationally. For example: there is a country inhabited by a masochistic people. Although they are a people, schooled in a tragic history, they haven't learned anything as a result. They aren't aware that a period of abundance, a miserable robotic policy, and anarchy do not last forever....and that will happen.

Where are the sane people with broad and futuristic vision?

HAIR LOSS

Old age doesn't begin at a certain age. We experience hardships from the moment we are born. Every living creature instinctively strives upwards as a precondition for survival. The hardships aren't symmetrical, as each person experiences difficulties in a different way. Since the "self" produces the bodily sensations, the body is carried and operated by its owner's sensations. It's a sensation of the "self" feeling bald. We can change it by using a diverting imagination: convincing ourselves that we suffer from hair too thick (the body can't tell between right and wrong, imagination and reality); or feeling that the rich head of hair is bothering us, since it is too rough and grows long too fast.

A reminder: don't look in the mirror when you deny a certain phenomenon. It should be mentioned that the sensations of fatigue, the sense of being burdened, the responsibility we bear and the character of men are all different from those of women.

As for body hair, it has no positive role. Men are advised to get rid of it, for esthetic reasons. One should not be blindly faithful to social conventions. Interpretations are allowed to be changed.

HYPNOSIS

I have mentioned throughout this book that hypnosis is a state where the hypnotized individual knowingly and passively consents to passing the "steering wheel" of his "self" to the hands of the hypnotizer, under the condition that he trusts him and commits to him in an absolute manner. An individual accumulates knowledge throughout his life, including imaginary knowledge, which he never actually experienced (by his individual interpretation). This information is out there in the subconscious, and there is no mechanism that purposely erases it for any kind of aim. Each individual uses this information based on his ability and mental state. In other words, the more the "self" feels it is drowning, the more the information fades and drifts farther away. You should keep in mind that the information doesn't disappear. That is, the conscious "self" is merely drawing the information from the subconscious "self," according to its individual ability at a given time.

The memory manifests not by will or need. **There is no entity directing us for our well-being**. The more accentuated the memory is, the more it surfaces, until it finally emerges.

For these reasons, the hypnotizer can plant information in

the hypnotized person's "self," or draw real or imaginary information from it. In this situation, the hypnotized body operates and reacts depending on the controlled and passive situation he is currently in, because the body is governed and operated by the hypnotized "self."

It's not the brain that is hypnotized. It is the infinite "individual self" thinking, using an unaware biological computer.

Comment: We should strictly forbid an unauthorized person from conducting such experiments on human beings. And even then, the aim must be completely justified. If the hypnotized person loses his way back to reality, we should repeat the action and show him the beginning of his path and the processes he needs to go through to get there. That person should be tempted, with small and guided steps, back to a more seductive and interesting reality, without force, rituals, or convincing; but only with reason and explaining the principle.

WITCHCRAFT

There is a mystery when the cause of a phenomena is unknown, yet when the principle is revealed, the mystery evaporates and turns into a meaningless word.

For example: a famous man lives in India who claims he has "x-ray" vision. He supposedly serves as a messenger and is able to see the body's internal organs, blood vessels, and bones.

An explanation: as we all know, the human senses are dimmed, faded, and lost. Animals, unlike humans, have highly sensitive senses that are aware of the body language of those surrounding them. It's a condition for survival. The predator "self" recognizes the weak and tired, and locates them through his imagination, which is like x-ray vision. The predator *"self"* actually identifies the internal weakness points of the prey's body language, by its interpretation, which is as sharp as a knife. The prey's "self" recognizes the predators by the same principle – a serious, strong, weak, hungry, satiated, sick, healthy, coward, pushed down, hurt, determined predator and so on. Without those senses, the prey would be in constant flight since birth, which is of course impossible. The senses of man are out there. Each man has a specific talent, usually hidden and wasted, that doesn't manifest itself. Sometimes, for some reasons, he discovers his talent and expresses it. And by that logic it is safe to assume that they are all, in fact, lost magicians. A demonstrative example: we hear through our ears the voices of others and interpret them as language by an individual interpretation. When there are noises in the background, we lose our connection to the man speaking to us. On the other hand, a person who is born deaf could interpret the silent language by lip reading, facial expression, and body language. Therefore, we discover that there are a lot of lost senses we are unaware of. That is the sunken person's "self." The brain is there.

Possession

It happens that the individual "self," which is caught in a dizzy spell, focuses on a specific character according to his belief, and then "sinks" and fades into it as it drifts away from reality. In this new situation, he acts according to that particular persona, by his individual interpretation. Sometimes he mumbles strange languages that are completely foreign to him.

This is information that he has accumulated without being aware (like, watching a foreign film with subtitles). When an individual sinks into a different persona, his realistic surroundings are no longer relevant, so all his abstract focus is limited to that event (see "**Conscious and Subconscious**" episode).

Don't let this supposed demon emerge, not through magic or through the fingernails, the belly button, or the buttocks. Only an authorized person must explain to him the path that he is on, and guide and lead him back into the tempting reality, using the right words and sentences. **The brain is completely operative.**

Acceptance

When the "self" believes something, any theory, according to his interpretation, becomes completely legitimate as far as he believes. That's according to the biological computer, the brain, which is lacking sense and independent cognition. Because of this, when a man is convinced by someone else of some fact, and is now a believer, according to his interpretation, even a fact such as $1+1=3$ can become legitimate and absolute; and it is no longer useful to argue with him on this point, since he is in the new and sealed individual maze. We should keep in mind that complex beliefs have developed from a belief in the first spark and from there on by interpretations over interpretations.

I should mention that the theories based on beliefs are thought of by logic. It's an individual logic, according to the interpretation of the believer, aimed at the believer, and should cross paths with the guidelines of the theory (in a subconscious way). That's the reason why many theories that are formed according to belief ease the mind. And those that don't are avoided or receive an indirect, "smart ass" interpretation. The brain cannot find out whether the interpretation is right or wrong.

This principle has caused, is causing, and will continue to cause accumulating tragedies.

CARRYING CHRONIC ILLNESSES

Let's say a child suffers from an eye infection, caused by reasons that I mentioned previously ("the quicksand swamp.") This illness is not contagious, yet other people may catch it.

If a father relates to the poor sensation of his son and senses: "I had better have the helpless toddler's misery passed on to me so it will be easier on him," or "I'm convinced that his illness is passing on to me." By then the doors to the disease are opened and he might get it. The distance between health and sickness is very small. Again, this is all according to the floating and fading limits of the "quicksand swamp." It's an internal, unconscious sensation, formed by an individual and subconscious interpretation. This principle is relevant to a variety of contagious illnesses.

A real case was documented on the National Geographic channel, where a poisonous spider bit a woman on the leg. The woman got extensive treatments, yet her condition continued to deteriorate. Her leg responded very severely to the venom. Her condition kept getting worse and no cure could be found to relieve it.

An explanation

It's a state in which that same woman carries with her the traumatic sensations she has yet to part with. In other words, she still tangibly feels that spider bite in her imagination, and is convinced that the venom is spreading, continuing to cause damage and eating up her leg. The past is, in fact, the present, as far as she is concerned. There goes the deadly chain of reactions, depending on a destructive individual interpretation (see "**Concept of Time**" episode).

The body has an absolute ability to clean the venom, even if is a black widow spider.

There are various chronic illnesses that carry traumas from the past with them. It is a state where the "self" senses a disease as a reaction to a subconscious situation from the past or present, or that one that will occur in the future, or any other memory that the individual interprets as traumatic.

This issue brings us back to the concept of "time." The time in this situation stands still. The sickly "self" is still there. The body is managed and operated by these sensations. Thereby the snowball phenomenon begins. When the disease is irritating, it becomes more real. When it is irritating and real, it becomes more tangible and frustrating. And so on, toward the abyss.

Therefore, we should be aware of the fact that it is a deliberate phenomenon of the body's owner, the "self." We are capable of completely curing ourselves of these phenomena.

There are illnesses that characterize a certain personality. Meaning, an individual is born and raised in a grim atmosphere. This sensation matches, for instance, skin diseases. This disease doesn't have an expiration date. As long as that individual is raised with the same point of view and the accompanying sensation (like: habits, body language, speech, attire, company, thought, atmosphere, etc)., the more chronic the disease becomes in his life, as it grows and matures alongside him, and is fashioned in line with the sickly childhood sensations. That means he carries it with him (see **"Concept of Time"** and **"Resistance"** episodes).

Recommendations

At the initial stage, the person must part ways with his "self pity card" and the chronically sickly "self."

At the second stage, one must be rough and rugged, with a fighter's personality, to allow him to shove his way out of each disturbing occurrence, just like wild animals do.

At the third stage, one should minimize his traumatic occurrences. We should all learn to leave them behind and shove our way forward, because life goes on. Don't remain stuck in a frozen situation from the past. The new interpretation should be: "This has happened to him or her. I am moving on."

Finally: We should activate our "resistance" instinctively.

As for that woman who was bitten by the spider. It's an example of people who enjoy illnesses and feeling unfortunate, without even being aware of it. They even interrupt other people in their surroundings, keeping them from continuing and surviving their lives.

SELF-CONFIDENCE

Self-confidence is a state in which the individual feels that a joyous occasion is coming along and that he is charged, and has an urge to realize it. This individual is convinced he can do it, with no stops or barriers along the way that could thwart the event. Meaning, there are stops and barriers, but they are not relevant. That's because, as **far as he is concerned, the event has already taken place, even if it is still in the planning stage and hasn't actually happened in reality. It's a kind of free conduction.**

That is, of course, an infinite individual state, that occurs within each person's "self," by their interpretation of that particular situation.

I should mention that blind self-confidence, without reasonable actions, leads to missing the point and could even cause harm.

The courses of life are determined and take place according to the determining "self." When an individual sees and "feels" obstacles that undermine and thwart his path toward his goal, he usually acts out the expected failure he foresees. Therefore, we should only focus on processes that advance the goal.

As for general lack of self-confidence, that individual feels naked and exposed in front of society. He sees himself through society's eyes, scanning and testing his awkward shortcomings. The more he focuses on a phenomenon, the more the situation is realized, similarly to the "scales principle." Instinctively, society takes advantage of his weakness and climbs all over him, while he is sinking down in a passive way, while others step on him, trampling him with delight.

Recommendations

We should take a couple of steps back, in order to view the grim situation in a wider and clearer perspective. We should collect energy, shove our way through, and enter society in a determined and relentless mood, as though we are setting the moves, since we all own life, without exception. It's now or never.

Since this principle is understood, it is possible to improve the quality of our lives by improving our self-confidence. This can be done when the individual teaches his "self" to cross barriers with "self-confidence" and achieve goals more easily.

The Invention that Changed our Lives

Imagine how life would look if man hadn't been able to discover and create transparent substances. In fact, **glass** changed our lives much more than we are aware of. If glass didn't exist, there would be no lights to illuminate and no transparent windows in automobiles. There wouldn't be any mirrors, we couldn't have built jet planes; there would be no telescopes to watch the stars; we wouldn't have made it to the moon; and most importantly, there would be no microscopes.

As a byproduct, man wouldn't have discovered the germs. In other words, man wouldn't live in constant fear of germs, and therefore wouldn't worry about cleaning and washing anything. He wouldn't compulsively chase cleanliness and his body wouldn't be sensitive to bacteria. Man would live in peace with the germs (just like undomesticated animals). In this world, the germs wouldn't have "lifted their heads high." Moreover, man would have continued eating as he did in earlier times, and wouldn't consider irrelevant things like calories or "healthy food." He would conduct his life in perfect balance with nature. Man would have been stronger and more surviving.

Moreover, there wouldn't be any chemical substances that undermine the balance among animals, plants, man and the environment. For example: the bees, nowadays, are in danger of extinction, due to environmental pollution and chemical substances that cause mutations, and the strengthening of parasites and illnesses. To clarify, I remind you that through the repeated human confiscation of honey, the bees are forced to work harder with the process of pollination and creation of more food, which causes anxiety, stress and exhaustion. This paves the way to various diseases, death and eventually extinction. Unlike what we thought, bees are not honey-making machines. The same goes for the production of fruit and vegetables.

Many plants are in danger of extinction, due to the fact that humans use chemical substances. Those plants are not resilient as they were before, and so they develop diseases that weren't created before. The same goes for animals. It's the connection between glass, bees, and hunger.

A Tip

A recommendation regarding the employment field.

Half the world works hard for most of their lives.

Half the world is unemployed.

The employees are exhausted due to overload.

The unemployed are worn-out by lack of productivity and poverty, and their ramifications (frustration, crime).

A law should be legislated: "week per week." Like a week for work and a week for rest. That way everybody will be occupied and life will be more balanced and less wearing out.

In order to suspend the process of annihilation, man must stop this blind (the dark maze) stupid and intolerable race of wasteful mass production and consumption. (Contrary to what people think, it will **save** the economy). Competition in manufacturing and industry must be qualitative and creative, not quantitative and mass. Implementation of this insight will indeed lead to a major cut in work, but will certainly enrich the state's treasury. For example, when a situation is created in which one person is forced to sell to someone else in a sly manner and the person's pocket is emptied. It is then that the stress begins and the person feels that they had "no choice". The result of this stress may be a sense of hopelessness, nervousness, aches, digestive discomfort, etc. One possibility is that he, in turn, will feel forced to cheat, steal, and harm others as well.

Another possibility is that he will be forced to take part in the unnecessary and wasteful animal industry (which has already been elaborated), whereas he is not at all aware of his participation in the extermination of animals. This is a distorted culture which humans have engaged in for generations and which has become acceptable and permanent (to be a partner in extermination, instinctively, thought and imagination – due to denial – are dulled and blocked. Now, he can no longer see the other from his perspective. As a result of this, there is a lack of awareness and its implications, which are the "dark maze"). A further possibility is that he is forced to take loans to manufacture unnecessary products that cause plunder, ruin, and annihilation. In the meantime, the factory equipment has worn out, and now, he needs to renew it. He experiences financial collapse, accompanied by losses. His country also loses the loan money and the imported equipment. His entire surroundings lose, which continues to repeat itself.

At the same time, we should spend more time on joyful and encouraging entertainment, replace gluttony culture with sports culture (without animals) making it a way of life and creating a new world order (according to this book). If the culture of atrophy continues, less successful children will be born in the next generation, to put it mildly. In the following generation, there will be a collapse.

PLANTS

It looks like plants are taken for granted. Never were the questions asked – who's in charge of their growth, and what lies beneath their development and growth? Well, plants have the "no choice," which is the "*Ultimate I am my own maker*." The plant's "self" lacks consciousness and emotions, but it definitely senses the "no choice" that is longing for energy, life, continuity and multiplication.

A detailed explanation will be given in the next book.

YOUTH RAGE

Here's a story of an unhappy child who frequently suffers violence from the religious teachers in the school he attends. Later on, his female dog and cat give birth during the same week, and both die from the same poison. The puppies remain orphans. The child then takes upon himself the responsibility of caring for them and nurturing them with milk. This takes a lot of time, causing him to be continuously late for morning prayers, and because of this he becomes subjected to sadistic reactions from his teachers. No more! The raging child goes out looking for a gun to kill the teachers, a number of bad individuals, and his sadistic father. He knows how to use a gun by watching TV. He knows that a gun contains only six bullets, and this fact, plus consideration for his mother, thwarts his plan because he wants to end all evil and himself as well; therefore, he needs much more than six bullets. But the rage keeps accompanying him until he drops out of school.

A concluding question: to what extent do we use our bodies and brain, or more precisely, not use our bodies and brain? The answer to this question is subjective.

This book was written with restraint and caution. Many paragraphs were censored. It is better that a man pulls together and operates according to the messages in this book.

I should salute the angels doing deeds for the salvation of the planet. They are doing sacred work. But they are too few. This is the sacred ideology, which must be spread around the world.

As for me, the writer, I have turned to specific sources, like associates and various scientists, in the country and overseas, claiming that I have new insights that contradict the rules and the norms. These references were made as the bare necessary minimum, and as could be expected, the comments were mocking from all directions. I was subjected to relentless, pitiful, and disgusting treatment, up until the last moment of this book's publication. After all, am I, the writer, supposed to add another book to the piles of books already written about health food and the like, with various and ridiculous studies and theories?

A message to my associates: it wasn't me. In fact, I was dressed in a kind of mask that says that everything is probable with me. Forgetting people, faces, names, events and incidents; and lacking the ability to communicate and the cognitive thinking. These were part of the senility (among others) that I was in, understandably at the moment.

Still there is a little secret. I am dyslexic. Among others, I have never written a letter. I am not educated – not in science or biology. I don't get along with a computer. I don't read many books and I have no education whatsoever. I recommend to all you readers out there, read this book twice so that you might cross the references and fully understand the content. I have insisted on this unique style of the book in order to set an example of a world belonging to everybody, while the big guys calling the shots up there aren't necessarily right.

This book will force the human being to face true reality as an existing fact that he can no longer ignore. Now the map of life is clearer and the puzzle of our existence is complete. For that reason, it is no longer possible for man to lose his way in reality.

In conclusion, the final sentence that should be accompanying each and every one of us during the course of our lives is: **Use your brain!** Why? Because it's there! In fact, everybody has a brain. You only need to open the door and step out of the "dark maze."

UPDATE

Last update: the year is 2010, a short time before publishing this book. I have gotten Flash back after many efforts and hardships. I will do whatever it takes for him to forget and forgive, if only a little, until I find him a new home.

The year is 2014,
The author of the book has disassociated himself from anything that belongs to and is reminiscent of the past. He has disassociated himself from the book and rejects it. He has not even read the book, not even to correct, complete

and edit it. He hardly remembers anything. Today, he is in the process of recreating himself. His diet consists of peanut butter sandwiches only. He will not write another book.

www.ingramcontent.com/pod-product-compliance
Lightning Source LLC
Chambersburg PA
CBHW051439170526
45166CB00001B/51